Spatial Demography Book Series

Volume 2

This series explores the application of spatial science to demographic information, including the methods, theory, and applications to societal problems. It captures the rapidly expanding knowledge-base of how human behavior and social processes are both shaped by place and time as well as change localities themselves over time. The books in this series examine both individuals as well as institutions and include all methodological, geographical, and topical research or treatments of the interaction between societies and place. It encompasses racial segregation, crime, urban sprawl, agricultural production, engagement in community life, migration, commuting, business location, technology, environmental quality, elections to public office, and other important societal phenomena. This series adopts an international and interdisciplinary approach in its detail of the tools, techniques, and theoretical perspectives used in spatial demography. The individual volumes will help demographers better understand when, how, and why space matters in social behavior and institutions.

More information about this series at http://www.springer.com/series/13093

David Darmofal • Ryan Strickler

Demography, Politics, and Partisan Polarization in the United States, 1828–2016

Springer

David Darmofal
Department of Political Science
University of South Carolina
Columbia, SC, USA

Ryan Strickler
Department of Political Science
Colorado State University Pueblo
Pueblo, CO, USA

Spatial Demography Book Series
ISBN 978-3-030-03999-8 ISBN 978-3-030-04001-7 (eBook)
https://doi.org/10.1007/978-3-030-04001-7

Library of Congress Control Number: 2018964105

This Springer imprint is published by the registered company Springer Nature Switzerland AG.
The registered company address is: Gewerbestrasse 11, 6330 Cham, Switzerland

Preface

In his keynote address at the 2004 Democratic National Convention, then-State Senator and soon-to-be US Senator Barack Obama rejected the argument that the United States was becoming a geographically polarized nation when he said "The pundits like to slice-and-dice our country into Red States and Blue States; Red States for Republicans, Blue States for Democrats."[1] The speech—and Obama's call for unity—would play critical roles in propelling him to the presidency just a little more than 4 years later. The appeal of Obama's call for unity across supposed lines of division and polarization was understandable. The appearance of the blue and red map on election night in 2000, and persistent references to it in the intervening years, had led Americans to wonder whether something had changed in the country, whether we had indeed become a geographically divided nation. These concerns were reinforced when, despite Obama's pleas to the contrary, the map was largely replicated in 2004. And a similar map appeared in 2008, in 2012, and in 2016.

Has something fundamentally changed in our American politics? Is this red state, blue state phenomenon both real and something new? Pundits have seemed to say yes but importantly have largely argued this by focusing only on the elections from 2000 to the present. These five presidential elections reflect only a small portion of America's electoral history. Indeed, they reflect only a little more than 10% of all presidential elections since the advent of mass voter participation in the 1828 contest between Andrew Jackson and John Quincy Adams.

We can understand our current era better, we believe, by placing it in historical perspective. By comparing current geographic patterns of voting to those in earlier eras, we can determine whether something has fundamentally changed in our politics. This was our motivation in writing this book, to place the current period of supposed geographic polarization in greater historical perspective by bringing to bear data from the entirety of the mass voter participation era in the United States, from 1828 to the present.

[1] Transcript at The American Presidency Project, http://www.presidency.ucsb.edu/ws/?pid=76988, accessed 3 February, 2018.

By placing our current period in historic relief, we found something that challenges many of the pundits and may provide solace to citizens concerned about their presumed increasingly divided nation. We've been here before and then some. The elections since 2000 don't stand out as unusually polarized elections, and indeed we've been much more geographically polarized in earlier eras, such as during the Jim Crow era of the Solid South. This finding, of course, comes as little surprise to political scientists and scholars in other disciplines who arguably have been more attuned to the historical timeline than popular observers.

Some of our other findings, however, likely will surprise these scholars, like the spatial structuring of voter participation that we uncover, for the first time, utilizing modern tools of spatial analysis. These tools allow us to identify where neighboring counties of similar Democratic or Republican (and before the Republicans, Whig) allegiances were located in each election. Or the fact that, contrary to the Big Sort argument, we aren't living in an anomalous period of geographic sorting. Or, perhaps most surprisingly, that knowing how a county voted in 1828 for Jackson or Adams is a better predictor of how it voted for Trump or Clinton in 2016 than is how it voted for Carter or Ford in 1976. Our book, in short, places our current period in historical relief and in doing so seeks to help us better understand our present moment as well as the past and how both are linked together in a geographic continuity that has marked our presidential elections in the United States.

We thank Chuck Finocchiaro, James Gimpel, Eitan Hersh, Nathan Kalmoe, Susan Miller, and Chris Witko for helpful conversations on this project and Luc Anselin and Wendy Tam Cho for helpful conversations on spatial analysis over the years. We also thank an anonymous reviewer for comments that have strengthened our book. We thank our editors at the Spatial Demography book series at Springer, Frank M. Howell and Jeremy R. Porter, and Bernadette Deelen-Mans for their help throughout the process of writing this book. We thank Peter F. Nardulli for the county-level data used in this book and the staff of the Rutherford B. Hayes Presidential Library in Fremont, Ohio.

Columbia, USA David Darmofal
Pueblo, USA Ryan Strickler

Contents

1 Introduction 1
1.1 Modern Partisan Polarization 2
1.2 Polarization and Geography 6
1.3 Outline of the Book 8
References 10

2 Beyond "Red State, Blue State": The Political Geography of Presidential Competition, 1828–2016 15
2.1 Partisanship, Voting, and Polarization 16
2.2 Polarization and Geography 17
2.3 The Political Geography of Presidential Voting 18
2.4 Analysis of County-Level Voting, 1828–2016 21
2.5 Political Geography and the American Party Systems 27
2.6 Conclusion 75
References 79

3 Geographic Polarization in Historical Perspective 83
3.1 Bishop's Argument 84
3.2 Recent Assessment of *The Big Sort* 85
3.3 Historical Examination of Geographic Polarization 88
3.4 Education and the Geography of Presidential Voting 93
3.5 Conclusion 95
References 98

4 Modeling the Political Geography of Presidential Voting 101
4.1 Diagnostics for Spatial Dependence 103
4.2 Analysis of Voting Behavior, 1828–2016 104
4.3 The Second American Party System 106
4.4 The Third American Party System 108
4.5 The Fourth American Party System 110
4.6 The Fifth American Party System 111
4.7 The Sixth American Party System 114

4.8 The Interaction of Racial and Immigrant Demographics 115
4.9 Conclusion 120
References 121

5 Clinton vs. Trump: Back to the Future? 123
5.1 The Relationship Between the 1828 and 2016 Elections............ 130
5.2 Looking Further on the 1828 and 2016 Relationship............ 132
5.3 Trump and Jackson: A Closer Look............ 133
5.4 Adams and Clinton: A Closer Look............ 136
References 140

6 Conclusion 143
6.1 Predicting the Future of Political Geography 146
References 148

List of Figures

Fig. 2.1 Within-state variation in partisan voting 22
Fig. 2.2 2016 Moran scatterplot 27
Fig. 2.3 The political geography of county-level voting in the 1828 presidential election 29
Fig. 2.4 The political geography of county-level voting in the 1832 presidential election 31
Fig. 2.5 The political geography of county-level voting in the 1836 presidential election 32
Fig. 2.6 The political geography of county-level voting in the 1840 presidential election 33
Fig. 2.7 The political geography of county-level voting in the 1844 presidential election 34
Fig. 2.8 The political geography of county-level voting in the 1848 presidential election 35
Fig. 2.9 The political geography of county-level voting in the 1852 presidential election 36
Fig. 2.10 The political geography of county-level voting in the 1856 presidential election 38
Fig. 2.11 The political geography of county-level voting in the 1860 presidential election 39
Fig. 2.12 The political geography of county-level voting in the 1864 presidential election 40
Fig. 2.13 The political geography of county-level voting in the 1868 presidential election 40
Fig. 2.14 The political geography of county-level voting in the 1872 presidential election 41
Fig. 2.15 The political geography of county-level voting in the 1876 presidential election 42
Fig. 2.16 The political geography of county-level voting in the 1880 presidential election 43

Fig. 2.17 The political geography of county-level voting in the 1884 presidential election 44
Fig. 2.18 The political geography of county-level voting in the 1888 presidential election 45
Fig. 2.19 The political geography of county-level voting in the 1892 presidential election 46
Fig. 2.20 The political geography of county-level voting in the 1896 presidential election 47
Fig. 2.21 The political geography of county-level voting in the 1900 presidential election 48
Fig. 2.22 The political geography of county-level voting in the 1904 presidential election 49
Fig. 2.23 The political geography of county-level voting in the 1908 presidential election 50
Fig. 2.24 The political geography of county-level voting in the 1912 presidential election 51
Fig. 2.25 The political geography of county-level voting in the 1916 presidential election 52
Fig. 2.26 The political geography of county-level voting in the 1920 presidential election 53
Fig. 2.27 The political geography of county-level voting in the 1924 presidential election 54
Fig. 2.28 The political geography of county-level voting in the 1928 presidential election 55
Fig. 2.29 The political geography of county-level voting in the 1932 presidential election 56
Fig. 2.30 The political geography of county-level voting in the 1936 presidential election 57
Fig. 2.31 The political geography of county-level voting in the 1940 presidential election 58
Fig. 2.32 The political geography of county-level voting in the 1944 presidential election 59
Fig. 2.33 The political geography of county-level voting in the 1948 presidential election 60
Fig. 2.34 The political geography of county-level voting in the 1952 presidential election 60
Fig. 2.35 The political geography of county-level voting in the 1956 presidential election 61
Fig. 2.36 The political geography of county-level voting in the 1960 presidential election 62
Fig. 2.37 The political geography of county-level voting in the 1964 presidential election 63
Fig. 2.38 The political geography of county-level voting in the 1968 presidential election 64

Fig. 2.39 The political geography of county-level voting in the 1972 presidential election 65
Fig. 2.40 The political geography of county-level voting in the 1976 presidential election 66
Fig. 2.41 The political geography of county-level voting in the 1980 presidential election 67
Fig. 2.42 The political geography of county-level voting in the 1984 presidential election 68
Fig. 2.43 The political geography of county-level voting in the 1988 presidential election 68
Fig. 2.44 The political geography of county-level voting in the 1992 presidential election 69
Fig. 2.45 The political geography of county-level voting in the 1996 presidential election 70
Fig. 2.46 The political geography of county-level voting in the 2000 presidential election 71
Fig. 2.47 The political geography of county-level voting in the 2004 presidential election 72
Fig. 2.48 The political geography of county-level voting in the 2008 presidential election 73
Fig. 2.49 The political geography of county-level voting in the 2012 presidential election 74
Fig. 2.50 The political geography of county-level voting in the 2016 presidential election 75

Fig. 3.1 Percentage of two-party voters living in landslide counties 89
Fig. 3.2 Percentage of two-party voters living in landslide counties, by party 89
Fig. 3.3 Map of landslide counties, 1976 90
Fig. 3.4 Map of landslide counties, 2016 91
Fig. 3.5 Map of landslide counties, 1828 91
Fig. 3.6 Map of landslide counties, 1904 92
Fig. 3.7 Percent over 25 with bachelor's degree, by party voting 94

Fig. 5.1 Presidential voting in Pike County, Kentucky 135
Fig. 5.2 Presidential voting in Hampshire County, Massachusetts 139

List of Tables

Table 2.1 Proportion of variance due to within- and between-state variation in margin of victory ... 23
Table 2.2 Global Moran's *I*'s: 1828–2016 ... 25
Table 2.3 Mean margins by Moran scatterplot category, 1828–2016 ... 76
Table 2.4 Percentage of counties in each Moran scatterplot category, 1828–2016 ... 77

Table 3.1 Correlation between county-level education and democratic vote share ... 94
Table 3.2 Percentage of voters living in "Landslide" counties, 1828–2016 ... 96
Table 3.3 "Landslide" counties, 1828–2016 ... 97

Table 4.1 Spatial models of county-level voting in the second party system, 1828–1852 ... 106
Table 4.2 Spatial models of county-level voting in the third party system, 1856–1880 ... 108
Table 4.3 Spatial models of county-level voting in the third party system, 1884–1892 ... 109
Table 4.4 Spatial models of county-level voting in the fourth party system, 1896–1928 ... 110
Table 4.5 Spatial models of county-level voting in the fourth party system, 1924–1928 ... 111
Table 4.6 Spatial models of county-level voting in the fifth party system, 1932–1956 ... 112
Table 4.7 Spatial models of county-level voting in the fifth party system, 1960–1964 ... 112
Table 4.8 Spatial models of county-level voting in the sixth party system, 1968–1992 ... 113
Table 4.9 Spatial models of county-level voting in the sixth party system, 1996–2016 ... 115

Table 4.10 Spatial models of county-level voting in the second party system, 1828–1852 ... 116
Table 4.11 Spatial models of county-level voting in the third party system, 1856–1880 ... 117
Table 4.12 Spatial models of county-level voting in the third party system, 1884–1892 ... 117
Table 4.13 Spatial models of county-level voting in the fourth party system, 1896–1920 ... 118
Table 4.14 Spatial models of county-level voting in the fourth party system, 1924–1928 ... 118
Table 4.15 Spatial models of county-level voting in the fifth party system, 1932–1956 ... 119
Table 4.16 Spatial models of county-level voting in the fifth party system, 1960–1964 ... 119
Table 4.17 Spatial models of county-level voting in the sixth party system, 1968–1992 ... 120
Table 4.18 Spatial models of county-level voting in the sixth party system, 1996–2016 ... 120

Table 5.1 Correlation with 2016 voting ... 131

Chapter 1
Introduction

Viewers tuning into TV coverage on election night 2000 found themselves confronting a new conception of the United States. A nation divided into two camps: one, a red country based in the South and the Plains and Mountain states and dominated by Republicans, and the other, a blue country located on the coasts and in the upper Midwest, dominated by Democrats. Had the map been a single election anomaly, or had it not seemed to capture something deeper, cultural, of which we were all becoming increasingly aware, this map would have been quickly forgotten, as previous maps from preceding elections had. But the map that viewers first encountered in 2000 largely reappeared in 2004, despite the occurrence of the worst terrorist attack on American soil and the resulting beginning of wars in Afghanistan and Iraq in the intervening years. And then it reappeared again in 2008, in a quite different election context featuring the first major party African American presidential candidate and taking place during the most severe economic crisis since the Great Depression. In 2012, in a country emerging from this crisis but marked by increased racial and political strife, it largely repeated itself. Even the surprising 2016 election, one argued to presage a new map realigned around populist politics, represented much more continuity than change, with nearly 90% of states (44 out of 50) voting as they had in 2012.[1] In all, 37 of the 50 states have voted for the same party in each of these five quite different election contexts. This pattern has seemed to capture deep-seated, divergent cultural predispositions that shape politics and many other aspects of our lives.

But is this pattern all that new? Although George Washington warned against "the baneful effects of the spirit of party" (Spalding 1996, 68), political parties and bitter partisan conflict have been a consistent feature throughout U.S. history.

[1]As marked by the winner of the statewide electors in the electoral college. One state, Maine, awarded its second congressional district elector to Donald Trump, the loser of its statewide contest.

D. Darmofal, R. Strickler, *Demography, Politics, and Partisan Polarization in the United States, 1828–2016*, Spatial Demography Book Series 2,
https://doi.org/10.1007/978-3-030-04001-7_1

And this conflict has long had a sharp geographic dimension. For example, long before the modern tension between the Democratic coast and Republican "flyover country," in the election of 1800 John Adams and the Federalists dominated the coast and Northern states, while Thomas Jefferson and the Democratic-Republicans dominated the South and the Appalachian region. A sharp divide in 1860 between the Republicans in the North and Democrats in the South precipitated the Civil War. The North/South partisan divide has persisted since; the only difference today is that the parties flipped after the Civil Rights Era, with the North now heavily Democratic and the South heavily Republican.

In sum, despite the handwringing over modern partisan polarization, it is not a new phenomenon. Using spatial analysis and a rich county-level dataset dating to the advent of Jacksonian Democracy, our book will explore the spatial and historical nuances of partisan geography, as well as the key factors that drive partisan voting behavior. While the insights presented are multi-faceted, two key themes will emerge—we have been here before, and the factors that drive partisanship today are very similar to those that have done so in the past.

1.1 Modern Partisan Polarization

For both popular pundits and scholars alike, the new, persistent, red and blue Americas seem to mark the potential emergence of a newly polarized America. In this new America, some argue, Democrats and Republicans have little in common ideologically, culturally, demographically, or geographically. Republicans see Democrats, as Jon Stewart opines, as "Marxists actively subverting our Constitution," and Democrats see Republicans as "racists and homophobes who see no one's humanity but their own" (quoted in Examiner.com 2010). As a result, David Brooks argues, politics has become nothing more than "good-versus-evil bloodsport" (2016). William Galston likewise laments that the inevitable result is "legislative gridlock and damage[d] trust and confidence in political institutions," causing "observers question America's ability to govern itself as the times require" (2014).

These pundits may be tapping into a real phenomenon, although the details are still in debate today. Researchers do agree that the ideological distance between partisan elected leaders is at a high water mark for the past century (Theriault 2008; Hill and Tausanovitch 2015). What's more, division across parties in Congress goes beyond ideology, spilling over to contentious debates on even valence and procedural issues (Lee 2009). A number of explanations are offered for this. McCarty, Poole, and Rosenthal attribute elite polarization, across history and today, to a rise in economic inequality and a subsequent move to the right by the Republican Party to prevent redistributionist policy (2006). As such, they claim that elite polarization is an asymmetrical phenomenon (see also Barber and McCarty 2013; Grossman and Hopkins 2016). Fiorina, however, points to a change in the party donor and activist base, where party careerists interested in patronage have been replaced by "purists" interested in ideological fidelity. This has only increased given the increasing demands campaign financing has placed on candidates in the twenty-first century (Lessig 2011). This explanation can also help account for, in contrast to the trends documented by McCarty and colleagues, Democrats' move to the left in the past few years (Campbell 2016); in 2012, for instance, the average member of Congress (both Democrats and Republicans) had to spend $10.2 million to win a Senate seat and $1.5 million to win a House seat. Increased fundraising demands create greater reliance on, and fidelity to, ideologically "pure" donors. Additional explanations for elite polarization focus on institutional features such as gerrymandering and direct primary elections, as well as a breakdown of informal norms of bipartisanship (for a discussion, see Barber and McCarty 2013, 23–35). Many, as well, ascribe a unique role to individual actors such as Newt Gingrich in explaining modern Congressional polarization and gridlock (Mann and Ornstein 2016, Ch. 2; Theriault and Rohde 2011).

Somewhat less clear, though, is the extent of polarization in the partisan *public*. Some such as Abramowitz (2010) claim that the electorate is splitting apart ideologically in much the same way partisan elites are. While much of the popular polarization narrative has focused on culture wars (Frank 2004), Abramowitz contends that Democrats and Republicans are farther apart on economic, cultural, and racial issues. In contrast, Fiorina, Abrams, and Pope contend that the electorate today is "closely" but not "deeply" divided (2010, 14). They argue that, once one digs past "red state" vs. "blue state" voting patterns, the public is ambivalent on even controversial issues such as abortion, gay marriage, and the war in Iraq. What's more, differences in opinion are minimal across parties and across racial, gender, and other demographic groups. Representing the majority view in the literature (Ansolabehere et al. 2006; Bafumi and Herron 2010; Levendusky et al. 2008; Fiorina 2009; Fiorina et al. 2010; Hill and Tausanovitch 2015), Levendusky (2009) likewise contends that the public is no more ideologically polarized than in the past; the partisan public is, though, "sorting," with Democrats increasingly more likely to adopt liberal positions and Republicans increasingly more likely to adopt conservative positions. This partly occurs through citizens choosing a "correct" party based on their beliefs, and partly by citizens shifting their beliefs based on their party attachment (Carsey and Layman 2006). Furthermore, recent research has explored the increasing social and affective polarization in the electorate. While differences in issue opinions may be relatively static, this work shows that partisans now have greater social distance than ever (Iyengar et al. 2012), and that Democrats and Republicans have increasingly sorted other group identity attachments—such as religious and racial identities—by their party identity (Mason 2016). This in turn fuels negative emotions such as anger towards one's outparty (Mason 2015), positive emotions such as pride and enthusiasm towards one's inparty (Huddy et al. 2015), implicit bias against outpartisans even on non-political matters (Iyengar and Westwood 2015), and less likelihood of engaging in open-minded reasoning with outpartisans (Strickler 2018).

Thus, the broad consensus that emerges from this literature is that, while the public is nearly as moderate on the issues as in the past, the *partisan* public is increasingly displaying an intra-party ideological consistency and interparty social and affective distance. Many posit that this is driven by the public's cuetaking from increasingly polarized elites. Both survey data analysis (Hetherington 2001; Darmofal 2005) and experimental research (Druckman et al. 2013) show that the public has increasingly been able to recognize interparty differences between politicians, pundits, and other elites and then adopt the attitude of their preferred party, even at the expense of following accurate information or expert opinion. The public also takes affective cues from party elites, as increasingly negative campaigning has driven an increase in negative outparty affect among partisans (Iyengar et al. 2012). Partisan public cuetaking, moreover, has only increased given structural changes in the media environment. With the rise of cable TV, internet news sites, and social media, a proliferation of information sources allows the modern consumer to engage in a process of "selective exposure," hewing to the partisan thought leaders that reconfirm their priors (Iyengar and Hahn 2009; Jamieson and Cappella 2008). While this allows those not interested to opt out of political media altogether, for interested partisans, it fuels greater ideological consistency through elite cuetaking (Prior 2007; Davis and Dunaway 2016).

Elite cuetaking and media changes go a long way to explain partisan sorting and social polarization. Demographic factors, however, also contribute to the public partisan split. The connection between partisan identification and demographic group identity is well established (Bentley 1908, 2008; Campbell et al. 1960; Green et al. 2002). Partisan conflict, thus, waxes and wanes with conflict among demographic groups. For example, the parties have long been divided on racial issues (Petrocik 1996), and the parties have also long been perceived to represent different races (Philpot 2007). There are numerous examples, further, of parties and platforms, explicitly or implicitly, motivated by anti-immigrant or anti-minority sentiment—the Know-Nothing Party of the 1850s, the Post Reconstruction Democratic "Solid South," and the party realignment of the late 1960s (fueled by white Southerner racial resentment) come to mind. Since the election of Barack Obama, though, polarized partisan politics has taken on a renewed racial/ethnic dimension. Today, partisan identification and issue opinions are much more cleanly divided along racial lines than even just before the start of his administration (Tesler 2016). Race-driven polarization can also be seen through the rise of the Tea Party; while the movement's origins and ideological commitments may be multifaceted, much of its energy and characteristic anger is driven by racially-tinged distrust of the Obama administration (Skocpol and Williamson 2012) as well as a general antipathy towards minority "Others" such as immigrants or the LGBT community (Parker and Barreto 2013). Racial polarization also can be seen with Donald Trump, whose nomination to the Republican Party was fueled by whites uneasy about demographic changes occurring in the United States (Tesler and Sides 2016). Beyond race and ethnicity, polarization today has come with citizens increasingly aligning with a party based on other identities, such as religious affiliation (Green et al. 2007) and sexual orientation (Frank 2012).

Real or merely perceived, elite and/or public, ideological or affective, popular punditry regularly decries the rise of polarization. David Brooks' lamentation of politics as "good vs. evil bloodsport" is more the rule than the exception; many in the academy, moreover, echo this negative take. One consequence of polarization many scholars point to is a lack of legislative productivity; despite admonishments to "get something done" and approval ratings dipping to a low of 7% in 2014 (Riffkin 2014), Congress remains mired in a higher level of legislative gridlock than it has experienced since the early twentieth century (Binder 2015). As expectations of bipartisanship and consensus have broken down, Sinclair (2008) has found that the quality of legislative deliberation in Congress has markedly decreased. Moreover, the number of public acts passed by the 2011–2012 Congress was the lowest it has been since the Civil War (Huder 2013). This policymaking lacunae, in turn, has fueled a rise in formal executive power, with executive orders, memoranda, agreements, and signing statements playing an increasingly important role in making policy (Mayer 2001; Lowande 2014). While perhaps necessary given Congress's inability to move the policy status quo, these tools raise serious concerns for the growth of presidential power and the circumvention of the Constitution's system of checks and balances.

In addition to issues with elite level politics, many point to normatively troubling consequences of polarization for the electorate. These scholars suggest that the consequences differ between independents and partisans. With modern polarization has come a rise of independents, eschewing an affiliation with either party (Pew Center 2015). Fiorina et al. (2010) argue that this is natural, as a largely moderate public is turned off by ideologically extreme, conflictual partisan rancor. While the evidence is somewhat mixed (see Abramowitz 2010), Rogowski (2014) finds that increased ideological distance between candidates does indeed depress voter turnout. For partisan identifiers, a long line of literature points to the fact that these individuals engage in motivated reasoning to support pre-existing partisan commitments (Kuklinski et al. 2001; Brady and Sniderman 1985; Lodge and Taber 2013; Gaines et al. 2007). This process of opinion formation led by affective group is only more prevalent in polarized political contexts (Druckman et al. 2013). What's more, partisans tend to predominately view news sources that only reconfirm their prior beliefs (Iyengar and Hahn 2009), a trend that has only increased in the age of social media (An et al. 2014).

Thus, in the age of polarization, partisan citizens increasingly consume the political world in a Democratic or Republican bubble, and they engage in this world with the implicit effort to support their party affiliation. As a result, just as in Congress, interparty public discussion and deliberation is increasingly likely to break down. Hetherington et al. (2016), for example, show a recent increase in "trait polarization;" partisans are less likely to view outparty Presidential candidates as merely wrong on the issues, but as personally flawed. This trait-based disdain "might cause people to support their side using tactics that produce a dysfunctional, gridlocked political system, as long as it hurt their opponents" (343). Indeed, Mutz (2006) finds that homogeneous interpersonal discussion networks decrease tolerance for oppositional viewpoints. Strickler (2018) finds that partisans are more likely to harbor anti-deliberative attitudes when confronted with outparty political disagreement; this is particularly true for those with strong social attachments to their party. The breakdown of deliberation, along with selective exposure to partisan media, means that the partisan public misses an opportunity to engage in learning and empathic learning. Thus, a key requirement for many democratic theorists—an informed citizenry and consensual, open-minded decision making—is on the decline in the modern polarized era.

Others, however, focus on more normatively positive consequences of polarization. For example, scholars have historically called for "responsible party government," where strong parties "are able to bring forth programs to which they commit themselves and possess sufficient internal cohesion to carry out these programs" (American Political Science Association Committee on Political Parties 1950, 1). And as theories of "conditional party government" suggest (Aldrich and Rohde 2001), elite-level polarization has gone hand in hand with stronger majority party control of Congress. While this may hamper inter-branch legislative coordination, within the House and the Senate strong parties set a clear and ideologically consistent legislative agenda. By doing so, they facilitate collective electoral accountability in two ways. For one, clear, and clearly opposing, party platforms provide clearer cues to voters, who are better able to vote based on their ideological preferences (Levendusky 2009). Partisans, moreover, are more likely to display "expressive" support for their party by donating, participating in a campaign, or trying to influence the vote of others (Huddy et al. 2015; Mason 2015). Given consistent concerns about voter competence (Converse 1964; Delli Carpini and Keeter 1996) and participation (Verba et al. 1995), this is no small benefit. In addition, evidence suggests that, with polarization, the majority party is more likely to be punished electorally for low overall Congressional approval ratings (Jones 2010). Although fighting against the headwinds of legislative gridlock, strong, polarized parties thus create an incentive to pass policy of which the public broadly approves.

Normatively "good" or "bad," inquiry into the causes and consequences of partisan polarization is now vital to comprehending the modern political landscape. Indeed, as the 2016 presidential election pitted two lightning rod candidates—loved by much of the party rank and file, but loathed by everyone else—against each other, understanding polarization will only grow more vital for the foreseeable future.

1.2 Polarization and Geography

With the public debate on polarization, geography has seen a good deal of attention. Commentators such as David Brooks (2016) and Thomas Frank (2004), for example, have noted that in recent elections "the old Democratic regions of the South and the Great Plains were on (the Republican's) team now, solid masses of uninterrupted red, while Democrats were restricted to the old-line, blue-blood states of the Northeast, along with the hedonist left coast" (Frank 2004, 14). In *The Big Sort*, Bill Bishop laments that "ways of life now have a distinct politics and a distinct geography" (2008, 14).

The public interest in this apparent geographic polarization has been matched by a renewed scholarly interest in geographic polarization and, more broadly, in political geography. A key recent example of this is Katherine Cramer's book *The Politics of Resentment* (2016). Using Wisconsin and the political ascendency of Scott Walker in 2010 as her extended example, she argues that the rural/urban divide that is a strong feature of the modern political landscape is driven, in part, by rural social identity and consciousness. Through extended interviews with small groups in communities across the state, she demonstrates how this consciousness drives a disdain for urban and liberal "elites" as well as a perspective that rural interests are ignored. Other recent examples include scholarship that uses voting data—at the state, county, zip code, or precinct level—to examine claims made by Bishop in *The Big Sort* (e.g. Abrams and Fiorina 2012; Myers 2013; Johnston et al. 2016; Martin and Webster 2017). In many ways, these works reflect a return to concerns that animated scholarship in earlier decades, before political behavior scholars started focusing predominantly on public opinion surveys of randomly selected respondents. In this earlier era, aggregate election returns were the principal data employed by scholars of political behavior. And geographic polarization was a principal concern of these researchers.

Influential scholars such as V.O. Key, Jr. (1949, 1955, 1959), E.E. Schattschneider (1960), and Walter Dean Burnham (1965, 1971) had called our attention to separate, solidly Democratic and Republican electorates with their attention on realignment theory and the 'system of 1896.' Realignment theory argued that enduring change was induced in geographically localized electorates in which emergent issues or crises led citizens to adjust their voting behavior and adopt enduring Democratic or Republican loyalties.[2] The 'system of 1896' thesis argued that the country had seemed to split into two separate countries after the 1896 election, a solid Democratic South and a solid Republican non-South (albeit one in which the parties' positions were often the diametric opposites of their positions during the current era of supposed geographic polarization).

The strength of analyses based on aggregate electoral results is that they place a primacy on space and time. That is, they highlight the importance of place and political geography—and of over-time changes in voting behavior—for our understanding of this behavior in the United States. As a consequence, they help us understand voting behavior in the geographically located local electorates in which we live our lives. They, in short, shed light on voting behavior in the real world, in the places where we live, much as the geographically-focused election maps since 2000 have as well. But in emphasizing aggregate electoral returns from cities, counties, and states, they necessarily are limited in what they can tell us about individual-level motivations for voting lest they fall into the ecological fallacy in

[2]Such change may be produced by three realignment dynamics: the conversion of active partisans from one party to the other, the demobilization of previously active voters, or the mobilization of previous non-voters. For further discussion and analysis of these realignment dynamics see Darmofal and Nardulli (2010).

which we attempt to infer behavior of the trees (voters) from data on the forests (the cities, counties, and states) in which they reside.[3] The post-World War II behavioral revolution placed a primacy on understanding these individual-level motivations. And as a consequence, political science, like its sister social science discipline of sociology, moved increasingly toward employing individual-level survey data to understand behavior.

The insights of survey research in political science are extensive, far too large to be catalogued here. But with the move away from aggregate electoral analyses, we also lost the two advantages of the earlier approach to understanding political behavior. A survey of a randomly selected 1000 or 1500 survey respondents can tell us much about individual-level motivations for voting behavior. But it necessarily loses much of the geographic dimension that gives life to politics. Many geographic locales are simply not sampled in standard public opinion surveys. And for those locales that are sampled, we often have too few survey respondents to say anything meaningful about the effects of their geographic context on their behavior.[4] And because modern, scientific surveys with random sampling date only to the 1940s, scholars lose much of the time span of mass political behavior in the United States in primarily employing survey data.

1.3 Outline of the Book

Our book is an effort to bring geography and history back into the study of partisan political behavior. It also seeks to add spatial and historical context to the debate on partisan polarization. While the insights provided here defy easy categorization, an overarching theme of the analysis and discussion that follows is that it is all too easy to overstate the anomaly of twenty-first century partisan politics.

In Chap. 2, we use presidential election data and spatial analysis to examine the geographic structure of partisan voting in U.S. history. Previous literature on geography and voting tends to examine patterns at the state or regional level; it also tends to focus only on recent decades. In contrast, we use county-level data to identify sub-state and trans-state regimes of partisan support from the start of the Second Party System in 1828 up to the 2016 election. After a review of global and local measures of spatial autocorrelation used, we go in detail for each election, matching statistical analysis and GIS mapping with an in-depth historical and contextual discussion. While each election does have its unique characteristics, we demonstrate that, across individual elections, continuity rather than change is the rule. Even with very different candidates, issues at play, and margins of victory,

[3]For further discussions of the ecological fallacy see Robinson (1950) and King (1997).

[4]Multi-level regression with poststratification, however, presents a fruitful approach for estimating subnational quantities of interest from national survey data (see, e.g., Gelman and Little 1997; Park et al. 2004; Lax and Phillips 2012).

within each historical party system the areas of partisan support remain relatively stable. This attests to the power of partisan identification in the American electorate and the slow nature of partisan geographic change.

Chapter 3 tackles the topic of geographic polarization, using Bill Bishop's influential book *The Big Sort* (2008) as a foil. Are we "sorting" into increasingly homogenous, increasingly opposed communities, as Bishop claims? A review of recent literature suggests that communities are perhaps becoming less competitive with regard to *recent* presidential election results; migration of individuals to "like minded communities," however, is not a primary cause of this. What's more, our analysis shows that recent geographic polarization has to be placed in historical context. While we are more polarized as compared to 1976, we are not as compared to 1828, or 1860, or 1904. What's more, the strongest evidence for geographic "sorting" based on lifestyle and culture comes only from 2016, the most recent election. Time is needed to tell if this unusual election is a blip, or the start of a historically significant pattern of polarization.

In Chap. 4, we seek to address what drives historical patterns of partisan geography, and why these patterns only change slowly over time. Employing spatial lag and spatial error models, we assess the role of race, immigration, and population density in driving county-level voting patterns in each election since 1828. We show that, in the vast majority of elections, spatial lag models are preferable to spatial error models. This implies that counties should not be seen as atomistic entities (only exhibiting a spatial structure due to shared partisan/demographic characteristics), but as exerting influence on one another through a process of diffusion. Moreover, while race, immigration, and population density have varying influences on voting behavior for each party system examined, across all party systems through history race plays a consistent and strong role in voting behavior. What's more, it is only from the 1920s on that population density has a significant effect on voting behavior. Thus, the urban/rural divide that plays such a defining role in politics today (Cramer 2016) is not a constant through history.

Chapter 5 offers a comparison of the geographic bases of partisan support between Andrew Jackson's election in 1828 and Donald Trump's election in 2016. Many commentators, including Trump himself, have drawn the comparison between these two elections. In this chapter, we review the key components of the Jacksonian populist tradition and how they apply to Trump. The pillars of "folk community" that Jackson extolled connect clearly to the brand of white identity politics, anti-immigrant sentiment, and fear of cultural displacement that motivated voters' support for Trump. What's more, a comparison of county-level voting from 1828 to 2016 shows that counties that voted strongly for Jackson tended to do so for Trump as well while counties that voted strongly for Adams tended to do likewise for Clinton. The chapter concludes with an in-depth discussion of two counties that exemplify the connection between these two elections—Pike County, Kentucky (a strong Jackson/Trump county), and Hampshire County, Massachusetts (a strong Adams/Clinton county).

Chapter 6 concludes our book. We examine the insights that have been generated by placing our current political period in historical context. And we consider what the future might hold for political polarization in the United States.

References

Abramowitz, A. I. (2010). *The disappearing center: Engaged citizens, polarization, and American democracy*. New Haven: Yale University Press.

Abrams, S. J., & Fiorina, M. P. (2012). The big sort that wasn't: A skeptical reexamination. *PS: Political Science and Politics, 45*(2), 203–10.

Aldrich, J. H., & Rohde, D. W. (2001). The logic of conditional party government: Revisiting the electoral connection. In L. C. Dodd & B. I. Oppenheimer (Eds.), *Congress reconsidered* (7th edn.). Washington, D.C.: CQ Press.

American Political Science Association Committee on Political Parties. (1950). Toward a more responsible two-party system. *American Political Science Review, 44*(3), 2: i–99.

An, J., Quercia, D., Cha, M., Gummadi, K., & Crowcroft, J. (2014). Sharing political news: The balancing act of intimacy and socialization in selective exposure. *EPJ Data Science, 3*(1), 1.

Ansolabehere, S., Rodden, J., & Snyder, J. M. (2006). Purple America. *The Journal of Economic Perspectives, 20*(2), 97–118.

Bafumi, J., & Herron, M. C. (2010). Leapfrog representation and extremism: A study of American voters and their members in congress. *American Political Science Review, 104*(03), 519–42.

Barber, M., & McCarty, N. (2013). Causes and consequences of polarization. In J. Mansbridge & C. J. Martin (Eds.), *Report of the task force on negotiating agreement in politics* (pp. 19–53). Washington, D.C.: American Political Science Association.

Bentley, A. F. (1908). *The process of government: A study of social pressures*. New Brunswick: Transaction.

Binder, S. (2015). The dysfunctional Congress. *Annual Review of Political Science, 18*, 85–101.

Bishop, B., with Cushing, R. G. (2008). *The Big Sort: Why the clustering of like-minded Americans is tearing us apart*. Boston: Mariner Books.

Brady, H. E., & Sniderman, P. M. (1985). Attitude attribution: A group basis for political reasoning. *American Political Science Review, 79*(4): 1061–78.

Brooks, D. (2016). How to fix politics. *New York Times*, April 12, 2016, Web edition. http://www.nytimes.com/2016/04/12/opinion/how-to-fix-politics.html?rref=collection%2Fcolumn%2Fdavid-brooks

Burnham, W. D. (1965). The changing shape of the American political universe. *American Political Science Review, 59*, 7–28.

Burnham, W. D. (1971). Critical elections and the mainsprings of American politics. New York: W.W. Norton & Company.

Campbell, J. (2016). *Polarized: Making sense of a divided America*. Princeton: Princeton University Press.

Campbell, A., Converse, P. E., Miller, W. E., & Stokes, D. E. (1960). *The American voter*. New York: Wiley.

Carsey, T. M., & Layman, G. C. (2006). Changing sides or changing minds? Party identification and policy preferences in the American electorate. *American Journal of Political Science, 50*(2), 464–77.

Converse, P. E. (1964). The nature of belief systems in mass publics. In D. E. Apter (Ed.), *Ideology and discontent* (pp. 206–61). New York: The Free Press.

Cramer, K. J. (2016). *The politics of resentment: Rural consciousness in Wisconsin and the rise of Scott Walker*. Chicago: University of Chicago Press.

Darmofal, D. (2005). Elite cues and citizen disagreement with expert opinion. *Political Research Quarterly, 58*(3), 381–395.

Darmofal, D., & P. F. Nardulli. (2010). The dynamics of critical realignments: An analysis across time and space. *Political Behavior, 32*(2): 255–283.

Davis, N. T., & Dunaway, J. L. (2016). Party polarization, media choice, and mass partisan-ideological sorting. *Public Opinion Quarterly, 80*(S1), 272–97.

Delli Carpini, M. X., & Keeter, S. (1996). *What Americans now about politics and why it matters.* New Haven: Yale University Press.

Druckman, J. N., Peterson, E., & Slothuus, R. (2013). How elite partisan polarization affects public opinion formation. *American Political Science Review, 107*(1), 57–79.

Fiorina, M. P. (2009). *Disconnect: The breakdown of representation in American politics.* Oklahoma: University of Oklahoma Press.

Fiorina, M. P., Abrams, S. J., & Pope, J. C. (2010). *Culture war? The myth of a polarized America* (3rd ed.). New York: Pearson.

Frank, B. (2012). Party polarization is now complete. *Gay & Lesbian Review Worldwide, 19*(5), 10–12.

Frank, T. (2004). *What's the matter with Kansas?: How conservatives won the heart of America.* New York: Metropolitan Books.

Gaines, B. J., Kuklinski, J. H., Quirk, P. J., Peyton, B., & Verkuilen, J. (2007). Same facts, different interpretations: Partisan motivation and opinion on Iraq. *Journal of Politics, 69*(4), 957–974.

Galston, W. A. (2014). Americans are as polarized as Washington. *Wall Street Journal.* June 3, 2014. http://online.wsj.com/articles/william-a-galston-americans-are-as-polarized-as-washington-1401837373

Gelman, A., & Little, T. C. (1997). Poststratification into many categories using hierarchical logistic regression. *Statistics Canada, 23*(2), 127–135.

Green, D. P., Palmquist, B., & Schickler, E. (2002). *Partisan hearts and minds: Political parties and the social identities of voters.* New Haven: Yale University Press.

Green, J. C., Kellstedt, L. A., Smidt, C. E., & Guth, J. L. (2007). How the faithful voted: Religious communities and the presidential vote. In D. E. Campbell (Ed.), *A matter of faith: Religion in the 2004 presidential election* (pp. 15–36). Washington, D.C.: Brookings Institution Press.

Grossman, M., & Hopkins, D. A. (2016). *Asymmetric politics: Ideological Repubicans and group interest Democrats.* Oxford: Oxford University Press.

Hetherington, M. J. (2001). Resurgent mass partisanship: The role of elite polarization. *American Political Science Review, 95*(3), 619–631.

Hetherington, M. J., Long, M. T., & Rudolph, T. J. (2016). Revisiting the myth: New evidence of a polarized electorate. *Public Opinion Quarterly, 80*(S1), 321–50.

Hill, S. J., & Tausanovitch, C. (2015). A disconnect in representation? Comparison of trends in congressional and public polarization. *Journal of Politics, 77*(4), 1058–75.

Huddy, L., Mason, L., & Aarøe, L. (2015). Expressive partisanship: Campaign involvement, political emotion, and partisan identity. *American Political Science Review, 109*(01), 1–17.

Huder, J. (2013). Our very unproductive Congress. Government Affairs Institute: Georgetown University. http://gai.georgetown.edu/our-very-unproductive-congress/

Iyengar, S., & Hahn, K. S. (2009). Red media, blue media: Evidence of ideological selectivity in media use. *Journal of Communication, 59*(1), 19–39.

Iyengar, S., Sood, G., & Lelkes, Y. (2012). Affect, not ideology: A social identity perspective on polarization. *Public Opinion Quarterly, 76*(3), 405–31.

Iyengar, S., & Westwood, S. J. (2015). Fear and loathing across party lines: New evidence on group polarization. *American Journal of Political Science, 59*(3), 690–707.

Jamieson, K. H, & Cappella, J. N. (2008). *Echo chamber: Rush Limbaugh and the conservative media establishment.* New York: Oxford University Press.

Johnston, R., Manley, D., & Jones, K. (2016). Spatial polarization of presidential voting in the United States, 1992–2012: The "big sort" revisited. *Annals of the American Association of Geographers, 106*(5), 1047–62.

Jones, D. R. (2010). Partisan polarization and congressional accountability in house elections. *American Journal of Political Science, 54*(2), 323–337.

Key, V. O., Jr. (1949). *Southern politics in state and nation*. New York: Knopf.

Key, V. O., Jr. (1955). A theory of critical elections. *Journal of Politics, 17*, 3–18.

Key, V. O., Jr. (1959). Secular realignment and the party system. *Journal of Politics, 21*, 198–210.

King, G. (1997). *A solution to the ecological inference problem: Reconstructing individual behavior from aggregate data*. Princeton: Princeton University Press.

Kuklinski, J. H., Quirk, P. J., Jerit, J., & Rich, R. F. (2001). The political environment and citizen competence. *American Journal of Political Science, 45*, 410–424.

Lax, J. R., & Phillips, J. H. (2012). The democratic deficit in the states. *American Journal of Political Science, 56*(1), 148–166.

Lee, F. E. (2009). *Beyond ideology: Politics, principles, and partisanship in the U.S. senate*. Chicago: University of Chicago Press.

Lessig, L. (2011). *Republic, lost: How money corrupts congress and a plan to stop it*. New York: Twelve/Hatchette Book Club.

Levendusky, M. (2009). *The partisan sort: How liberals became Democrats and conservatives became Republicans*. Chicago: University of Chicago Press.

Levendusky, M. S., Pope, J. C., & Jackman, S. (2008). Measuring district-level partisanship with implications for the analysis of US elections. *Journal of Politics, 70*, 736–53.

Lodge, M., & Taber, C. S. (2013). *The rationalizing voter*. Cambridge: Cambridge University Press.

Lowande, K. S. (2014). The contemporary presidency after the orders: Presidential memoranda and unilateral action. *Presidential Studies Quarterly, 44*(4), 724–741.

Mann, T. E., & Ornstein, N. J. (2016). *It's even worse than it looks* (2nd edn.). New York: Basic Books.

Martin, G. J., & Webster, S. (2017). Does residential sorting explain geographic polarization? Unpublished manuscript. http://polisci.emory.edu/faculty/gjmart2/papers/partisan_sorting_density.pdf

Mason, L. (2015). I disrespectfully agree. *American Journal of Political Science, 59*, 128–45.

Mason, L. (2016). A cross-cutting calm: How social sorting drives affective polarization. *Public Opinion Quarterly, 80*(S1), 351–77.

Mayer, K. R. (2001). *With the stroke of a pen: Executive orders and presidential power*. Princeton: Princeton University Press.

McCarty, N., Poole, K. T., & Rosenthal, H. (2006). *Polarized America: The dance of ideology and riches*. Cambridge: MIT Press.

Mutz, D. C. (2006). *Hearing the other side: Deliberative versus participatory democracy*. Cambridge: Cambridge University Press.

Myers, A. S. (2013). Secular geographical polarization in the American South: The case of Texas, 1996–2010. *Electoral Studies, 32*(1), 48–62.

Park, D. K., Gelman, A., & Bafumi, J. (2004). Bayesian multilevel estimation with poststratification: State-level estimates from national polls. *Political Analysis, 12*, 375–385.

Parker, C. S., & Barreto, M. A. (2013). *Change they can't believe in: The tea party and reactionary politics in America*. Princeton: Princeton University Press.

Petrocik, J. (1996). Issue ownership in presidential elections, with a 1980 case study. *American Journal of Political Science, 40*, 825–50.

Pew Center. (2015). A deep dive into party affiliation. http://www.people-press.org/2015/04/07/a-deep-dive-into-party-affiliation/

Philpot, T. (2007). *Race, Republicans, and the return to the party of Lincoln*. Ann Arbor: University of Michigan Press.

Prior, M. (2007). *Post-broadcast democracy: How media choice increases inequality in political involvement and polarizes elections*. Cambridge: Cambridge University Press.

Riffkin, R. (2014). Public faith in Congress falls again, hits historic low. Gallup. http://www.gallup.com/poll/171710/public-faith-congress-falls-again-hits-historic-low.aspx

Robinson, W. S. (1950). Ecological correlations and the behavior of individuals. *American Sociological Review, 15*(3), 351–357.

Rogowski, J. C. (2014). Electoral choice, ideological conflict, and political participation. *American Journal of Political Science, 58*(2), 479–94.

Schattschneider, E. E. (1960). *The semisovereign people: A realist's view of democracy in America.* New York: Holt, Rinehart and Winston

Sinclair, B. (2008). Spoiling the Sausages? How a Polarized Congress Deliberates and Legislates. In P. S. Nivola & D. W. Brady (Eds.), *Red and blue nation?* (pp. 55–87). Washington, DC: Brookings Institution Press.

Skocpol, T., & Williamson, V. (2012). *The tea party and the remaking of Republican conservatism.* New York: Oxford University Press.

Spalding, M. (1996). George Washington's farewell address. *The Wilson Quarterly, 20*(4), 65–71.

Strickler, R. (2018). Deliberate with the enemy? Polarization, social identity, and attitudes toward disagreement. *Political Research Quarterly, 71*(1), 3–18.

Tesler, M. (2016). *Post-racial or most-racial? Race and politics in the Obama era.* Chicago: University of Chicago Press.

Tesler, M., & Sides, J. (2016). How political science helps explain the rise of Donald Trump: The role of white identity and grievances. *Washington Post*, 3 Mar 2016, sec. Monkey Cage. https://www.washingtonpost.com/news/monkey-cage/wp/2016/03/03/how-political-science-helps-explain-the-rise-of-trump-the-role-of-white-identity-and-grievances/?tid=a_inl

Theriault, S. M. (2008). *Party polarization in congress.* Cambridge: Cambridge University Press.

Theriault, S. M., & Rohde, D. W. (2011). The Gingrich senators and party polarization in the U.S. senate. *Journal of Politics, 73*(04), 1011–24.

Verba, S., Schlozman, K. L., & Brady, H. E. (1995). *Voice and equality: Civic volunteerism and American life.* Cambridge: Harvard University Press.

Chapter 2
Beyond "Red State, Blue State": The Political Geography of Presidential Competition, 1828–2016

Both popular observers and scholars alike have become increasingly interested in the political geography of partisan voting in the United States. From the familiar "red and blue" maps that have become a focus of presidential elections since 2000 to both popular and scholarly studies of partisan sorting and the possibility of a "Big Sort" occurring, the political geography of voting is perhaps more prominent than at any time since V.O. Key was producing his seminal works on realignments (1955, 1959) and Southern politics (1949).

And yet while scholars are increasingly interested in examining and explaining the political geography of voting in the United States, we still know surprisingly little about the geographic details of this voting (but see Gimpel and Schuknecht 2003). Scholars have identified which states tend to vote Republican and which tend to vote Democratic, but a focus on substate patterns has been more limited, especially in historical context. And despite the focus on political geography in this literature, scholars have generally not employed methods of spatial analysis to document spatial regimes in partisan voting.

This chapter moves beyond existing studies in employing spatial analytical methods to identify where areas of Democratic and Republican strength have been located in presidential elections. It also moves beyond the contemporary focus of much of the existing literature by examining this spatial structure in all presidential elections since the advent of Jacksonian democracy in 1828. The analysis documents the changing—and unchanging—political geography of presidential elections in the United States.

This chapter is structured as follows. First we discuss our current understanding of partisanship and political geography. Next, we discuss global and local measures of spatial autocorrelation. In the following section we apply these diagnostics to presidential voting from 1828 through 2016. We conclude by discussing the implications of our analysis for understanding the political geography of presidential voting in the United States.

D. Darmofal, R. Strickler, *Demography, Politics, and Partisan Polarization in the United States, 1828–2016*, Spatial Demography Book Series 2,
https://doi.org/10.1007/978-3-030-04001-7_2

2.1 Partisanship, Voting, and Polarization

The relationship between partisanship and voting behavior has long been of interest to pundits and political scholars alike. Since *The American Voter*, many have suggested that party identification is a strong "psychological tie", an "unmoved mover" that exerts a strong, long-term influence on voting, as well as opinion formation and other political behavior (Campbell et al. 1960; Bartels 2002; Green et al. 2002; Carsey and Layman 2006). Others, however, suggest that short term factors—such as candidate assessments, retrospective evaluations, issue proximity, and the state of the economy or global affairs—play a stronger role in influencing vote choices than those positing party as an "unmoved mover" assume (Franklin and Jackson 1983; Fiorina 1981; Healy and Malhotra 2013; Jessee 2009). The effect of partisanship on candidate evaluation and vote choice, moreover, can vary by an individual's level of sophistication (Lau and David 2001) or ambivalence towards the parties (Basinger and Lavine 2005). In addition, the strength of partisanship in the electorate has waxed and waned over time. Through the 1970s and 1980s, scholars and pundits expressed considerable concern about the decline, or "dealignment" of parties in the United States electorate (Nie et al. 1979). Since the 1990s, however, many scholars assert that partisanship is at a high water mark not seen in a half century, and long-term party attachments have as strong an impact on voting behavior as ever (Bafumi and Shapiro 2009; Bartels 2000; Hetherington 2001; Shaw 2012).

Indeed, many suggest that, following the increasing ideological divergence of elites (McCarty et al. 2006; Theriault 2008), the public is now "polarized" along party lines. In other words, as the electorate has become more likely to identify with a political party than during the "dealignment" era, some have argued that the electorate is increasingly ideologically split, with Democrats and Republicans increasingly harboring homogeneous, distinct worldviews and issue positions (Abramowitz 2010; Abramowitz and Saunders 2008; Campbell 2008). Others have questioned this assertion. Fiorina, Abrams, and Pope, for example, contend that Americans remain "closely but not deeply divided," and no more ideologically driven than when *The American Voter* was written (2005, p. 13). Instead, they argue that the public is, to a certain degree, "sorting" into the "correct" party without significantly changing ideological dispositions (see also Levendusky 2009). Relatedly, there is debate as to what issues or dimensions partisans are "polarizing" on, if at all; for example, despite popular rhetoric concerning "culture wars" (Frank 2004), Gelman et al. (2008) assert that social issues have the highest explanatory power only for the voting of wealthier Americans, while low-income Americans are guided more by economic issues. Similarly, Carsey and Layman (2002) suggest that, while political elites have polarized on economic, cultural, and racial issues, the public has not uniformly done so. According to these authors, only the most politically sophisticated have picked up elite cues and polarized, while the rest of the public remains moderate on these issues.

2.2 Polarization and Geography

One prominent variant of the debate on polarization, public opinion, and voting behavior is whether America is exhibiting geographic polarization, captured in popular imagination by the divide between "red" and "blue" states or counties. Many argue that geographic polarization is more hype than reality. For example, Ansolabehere et al. (2006) as well as Fiorina et al. (2010) point to NES data showing that respondents in "red" states are very similar in ideological self-placement and opinions on issues to respondents in "blue" states (across both economic and social issues). Likewise, Evans (2003) finds that, while ideological and issue attitudes among Democratic and Republican identifiers have diverged, political attitudes across geographic regions have actually converged. Nivola and Galston (2008, 236), however, point to bitterly contested primaries and the decline of split ticket voting to suggest that the electorate is "clustering in 'red' and 'blue' counties, if not states or regions," (236). This argument is picked up by Johnston et al. (2016), who find increasing partisan geographic segregation over the past 40 years, leading up to the 2012 election, whether one looks at the state, county, or census tract level.

In *The Big Sort*, Bishop (2008) wades into the debate on geographic polarization, offering a provocative thesis on the increased "clustering" of like-minded Americans (see Chap. 3 for more detail). He uses a variety of opinion data to support the "sorting" thesis, but the most prominent pieces of evidence are county-level voting behavior since WWII (and in particular, from 1976 to 2004). For example, Bishop shows that, in presidential election races, the percentage of voters living in "landslide" counties, or counties where one candidate won by 20 percentage points or more, have increased from one quarter to one half since 1976. What's more, the number of "tipped" counties, or counties that consistently voted for one party for president for decades, has increased since 1948, and two-thirds of counties have become less competitive in presidential elections since 1976 (2008, Ch. 2). Bishop argues that this polarization of vote patterns both is caused by, and reinforces, a larger geographic polarization of American culture, where "pockets of like-minded citizens ... have become so ideologically inbred that we don't know, can't understand, and can barely conceive of "those people" who live just a few miles away" (2008, 40).

While much of the previous literature has examined the descriptive question of whether America is "polarizing" or "balkanizing," there has been comparatively less research into the causes of geographic polarization. Bishop, for example, offers the provocative, perhaps plausible, demography thesis: Americans have engaged in a "post-materialist Tiebout migration based on non-economic goods, as people have sought out places that best fit their ways of life, their values, and their politics," (2008, 199). However, the evidence he provides to support this causal story is largely anecdotal, and there is an opportunity missed to rigorously incorporate demography and migration studies into the typical explanations of partisan polarization promoted by political scientists. For example, internal migration of the electorate could have a dramatic effect on vote patterns, be it through the "compositional" shuffling of Democrats and Republican across localities (Gimpel and Schuknecht 2003; Jurjevich and Plane 2012), or the "contextual" partisan influence moving to a new location has on a prospective voter (Brown 1988). International migration could also affect the partisan landscape, as immigrant gateway regions (Frey 1999) provide large pools of "politically non-immunized" (Andersen 1979) first (and subsequent) generation citizens that strong local party organizations can convert and mobilize. Moreover, Stoker and Jennings illuminate the political effects of age demographics; as populations age, partisan and opinion attachment increase with age. Citizens in different age cohorts attach to the issues that were politically salient when they were young adults; thus, different cohorts have different salient issues (2008).

In addition, scholarship on geographic polarization often offers a blunt view of polarization, examining "red state" versus "blue state" dynamics (Fiorina et al. 2010; Abramowitz 2010; Abramowitz and Saunders 2008; Gelman et al. 2008). Other works examine a more fine-grained level, such as the county level of analysis (McKee and Teigen 2009; McGhee and Krimm 2009; Bishop 2008; Nivola and Galston 2008; Morrill et al. 2007). However, in these works, counties are treated as atomized, independent entities. Very little work takes into account spatial dependence: the existence of similar partisan behaviors for neighboring counties. Finally, most studies of geographic partisan polarization focus on the post WWII period—perhaps a unique time of partisan detente. Fischer and Mattson (2009), for example, argue, "Little in the contemporary scene matches the virulence of nineteenth- and early twentieth-century battles over alcohol, slavery, immigration, black enfranchisement, and compulsory schooling, which left property destroyed and bodies in the streets. Nor does contemporary conflict approach the turmoil of the mass demonstrations and political assassinations of the 1960s. And all these pale, of course, before the carnage of the Civil War, the specter that haunts all claims that America is falling apart" (436). While some work has taken a longer historical perspective (for example, see Gerring 1998), more can be done to connect the causes and complexities of past eras of partisan polarization to the modern era.

2.3 The Political Geography of Presidential Voting

At the heart of the interest by pundits in "red states" and "blue states" is an interest in the spatial structure of presidential voting. How is the political geography of voting structured? Where are areas of Republican strength and Democratic strength located?

This interest in the geographic structure of voting requires spatial methods to identify the political geography of voting, as our minds have a tendency to see patterns where none exist, a phenomenon known as apophenia. As a consequence, we can't simply look at maps of levels of partisan voting and identify the political geography of voting. Even if partisan voting were spatially random, we would be inclined to see spatial regimes of Democratic and Republican voting where none exist and conclude that there is a political geography of voting. Instead, we need to bring diagnostic tools to this question, to identify this structure and map it.

In this chapter we do this. We do something that scholars haven't done before: apply modern tools of spatial analysis to identify the political geography of voting from the beginning of mass voter participation in the United States in the 1820s through the 2016 election.[1] By employing a spatial diagnostic at the global level (specifically the global Moran's I), we diagnose whether there are spatial patterns to partisan voting in presidential elections, or whether partisan voting is distributed randomly across the country. We then move beyond the global level to probe this spatial dependence at the local level. By employing a local indicator of spatial association (LISA) statistic (specifically, the local Moran's I), we diagnose where spatial regimes of Democratic and Republican support have been located. Before presenting our diagnosis of spatial dependence in presidential voting, we first discuss both these global and local measures of spatial autocorrelation.

2.3.1 Global Measures of Spatial Autocorrelation

Global tests for spatial autocorrelation test for whether the data as a whole exhibit a spatial structure. For our analysis, does knowing where a county is located help us to predict its presidential margin (the Democratic share of the vote minus the Republican share of the vote)? Or are these margins randomly distributed with regard to space? If we identify a spatial structure at the global level, we can then turn to local spatial diagnostics to determine which counties share similar (or dissimilar) vote margins with their neighboring counties and which exhibit spatial randomness.

Why would we be better able to predict a county's margin of victory by knowing its location? In short, this would be the case if a county's margin of victory is spatially autocorrelated with the margins in neighboring counties. Formally, such spatial autocorrelation or spatial dependence (we use the terms interchangeably) reflects a non-zero covariance in the values on the variable at neighboring locations:

$$\mathrm{Cov}(y_i, y_j) = E(y_i y_j) - E(y_i)E(y_j) \neq 0 \quad \text{for} \quad i \neq j, \tag{2.1}$$

where i and j denote spatial locations (Anselin and Bera 1998, 241–242). Spatial autocorrelation (at either the global or local level) can take either of two forms. When positive spatial dependence is present, similar values on the random variable cluster at neighboring locations. When negative spatial dependence is present, dissimilar values cluster. In most applications, including this chapter's, we expect positive spatial autocorrelation to be more likely than negative spatial autocorrelation.

A critical term we have employed thus far, and one that needs further explicating, is the concept of "neighboring locations." The choice of the neighbors of unit i is a critical component in any spatial analysis, as it delimits the spatial autocorrelation that can be diagnosed and subsequently modeled. Neighbors are those units that may exhibit first-order spatial dependence with unit i (non-neighbors exhibit only

[1]See Darmofal (2015) for an extensive discussion of how to diagnose and model spatial dependence.

higher-order dependence with i, through neighbors of neighbors). In practice, the definition of neighbors is flexible and thus the tests for spatial dependence discussed here are also flexible.

In this analysis, we employ contiguous neighbors of county i as the neighbors of i. We do so for a simple reason that follows the conceptual interest in regimes of Democratic and Republican support. We're interested in where continuous spatial regimes of Democratic and Republican counties are located within—and across—states. For such a research interest, a queen contiguity neighbor definition in which all counties directly contiguous to i are neighbors of i is the natural choice of neighbor definition.

Whatever the researcher's choice of neighbor definition, this definition is reflected in a spatial weights matrix, W. In this matrix, neighbors of unit i have non-zero values, non-neighbors of i have zero values, and, typically (and in our analysis), i is treated as a non-neighbor of itself. It is common to row-standardize the spatial weights matrix so that the sum of weights for i's neighbors equals 1 and we follow this convention. This ensures comparability of spatial dependence estimates across tests and facilitates the interpretation of these estimates as correlations.

One of the principal tests for global spatial autocorrelation, and the one we employ in this chapter, is the global Moran's I. The global Moran test is:

$$I = \frac{N}{S} \frac{\Sigma_i \Sigma_j w_{ij}(y_i - \overline{y})(y_j - \overline{y})}{\Sigma_i (y_i - \overline{y})^2}, \tag{2.2}$$

with N denoting the number of observations, S the sum of the weights, w_{ij} an element of the spatial weights matrix W, y_i and y_j the values on the random variable at locations i and j, and $\overline{y}$ the mean on y.

A positive and significant global Moran's I indicates positive spatial dependence. High values cluster with high values at neighboring locations and low values with low values. In our analysis, this would mean that more Democratic counties cluster with neighboring Democratic counties while Republican counties cluster with neighboring Republican counties. A negative and significant global Moran's I indicates negative spatial autocorrelation, with high values bordering low values and vice versa. In our analysis, negative spatial autocorrelation would mean that strongly Democratic counties border strongly Republican counties and vice versa.

Importantly, a null finding on the global Moran's I does not necessarily imply the absence of spatial autocorrelation at the local level. Offsetting local spatial dependence, with positive autocorrelation prevalent in a portion of the spatial plane and negative autocorrelation prevalent in another could produce a null global result. As a result, even if the global Moran's I finds no evidence of global spatial autocorrelation, it is still important to examine autocorrelation at the local level to determine which units are exhibiting spatial dependence with their neighbors and the form (positive vs. negative) of this dependence.

Such local spatial dependence can be examined via the local Moran's I. This local Moran test is a member of a class of statistics known as LISA statistics: local indicators of spatial association. LISA statistics are local spatial autocorrelation tests that meet two conditions: (1) they identify which units are autocorrelated with their neighbors and, (2) their sum is proportional to a corresponding global measure (Anselin 1995). The local Moran's I is proportional to the global measure. Its use in connection with the global Moran's I allows us to disaggregate the global measure by identifying which units are producing the global finding, which run counter to it, and which exhibit spatial randomness.

The local Moran's I takes the form:

$$I_i = \frac{\Sigma_j w_{ij}(y_i - \overline{y})(y_j - \overline{y})}{(y_i - \overline{y})^2} \tag{2.3}$$

where the notation in (2.3) is identical to that presented for the global Moran's I in (2.2). A critical distinction between the local and global Moran statistics is that in the former, only the neighbors of i are incorporated in the separate estimates at each location while in the latter, summation occurs over all i.

Inference on the global and local Moran tests takes either of two approaches. If the variable of interest is normally distributed with a constant variance, then the Moran's I statistic is asymptotically normally distributed under the null. Inference then occurs by comparing an observed z-value to the probability of observing this if the null were true assuming a normal distribution.

Alternatively, if the researcher does not wish to assume normality, she can employ a permutation approach. In such an approach, the values on the variable of interest are randomly permuted across units to reflect the null hypothesis of spatial randomness. A Moran's I value is then calculated for each permutation, creating an empirical reference distribution against which the observed Moran's I is compared. In this chapter, we employ the permutation approach. Specifically, we randomly permute the values on our county-level measures of partisan voting 999 times to create a reference distribution.

2.4 Analysis of County-Level Voting, 1828–2016

Our measure for our analysis is the margin of difference in the presidential vote between the Democratic and Republican candidates at the county level (the Democratic proportion of the vote in the county minus the Republican proportion of the vote in the county). In other words, we are focused in this analysis on competition between the two major parties. We examine the spatial structure of this vote in all presidential elections from the beginning of mass voter participation in 1828 through the 2016 election. Before turning to our spatial analysis, we examine whether the maps of blue states and red states that have become commonplace in both popular punditry and academic analysis since 2000 accurately capture the spatial pattern of voting in the United States.[2]

[2]The data employed in this book are part of a county-level and state-level political, electoral, and demographic archive collected by Peter F. Nardulli and a team of researchers at the University

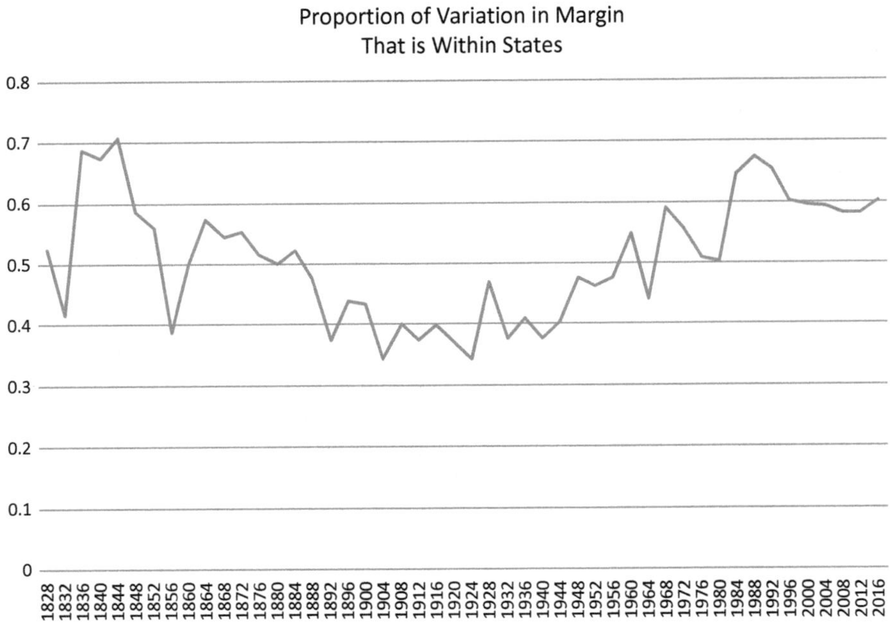

Fig. 2.1 Within-state variation in partisan voting

One way to examine this is to compare the variation in voting across states versus within states. At the extreme, if all variation in aggregate voting occurs between states as opposed to within states, then we will do quite well by simply comparing blue states vs. red states. Of course, this is largely a strawperson argument, as even pundits and scholars interested in blue states and red states don’t necessarily assume that within-state variation in voting is nonexistent. The focus on state-level variation, however, does imply that variation in voting across states is much greater and thus more consequential than variation within states.

Does between-state variation swamp within-state variation in voting? We examined this for each of the presidential elections in our analysis. Figure 2.1 presents the proportion of the combined within- and between-state variation in presidential vote margins that is produced by within-state variation. As we can see, within-state variation has accounted for a considerable portion of the variation in voting in each presidential election since the advent of Jacksonian democracy. On average, 51% of the variation is due to within-state variation. Within-state variation in voting has been just as widespread in American electoral history as between-state variation. Maps of blue states and red states literally tell only half of the story.

Table 2.1 presents the relative contributions of within- and between-state variation in partisan voting by election. The relative contribution of within- and between-state variation has varied over time. Variation in voting within states

of Illinois at Urbana-Champaign. The data include observations on all counties in the continental United States for each presidential election from 1828 through 2016.

Table 2.1 Proportion of variance due to within- and between-state variation in margin of victory

Year	Within-state	Between-state
1828	0.52	0.48
1832	0.42	0.58
1836	0.69	0.31
1840	0.67	0.33
1844	0.71	0.29
1848	0.59	0.41
1852	0.56	0.44
1856	0.39	0.61
1860	0.50	0.50
1864	0.57	0.43
1868	0.54	0.46
1872	0.55	0.45
1876	0.52	0.48
1880	0.50	0.50
1884	0.52	0.48
1888	0.48	0.52
1892	0.38	0.62
1896	0.44	0.56
1900	0.43	0.57
1904	0.34	0.66
1908	0.40	0.60
1912	0.38	0.62
1916	0.40	0.60
1920	0.37	0.63
1924	0.34	0.66
1928	0.47	0.53
1932	0.38	0.62
1936	0.41	0.59
1940	0.38	0.62
1944	0.40	0.60
1948	0.48	0.52
1952	0.46	0.54
1956	0.48	0.52
1960	0.55	0.45
1964	0.44	0.56
1968	0.59	0.41
1972	0.56	0.44
1976	0.51	0.49
1980	0.50	0.50
1984	0.64	0.36
1988	0.67	0.33
1992	0.65	0.35

(continued)

Table 2.1 (continued)

Year	Within-state	Between-state
1996	0.60	0.40
2000	0.59	0.41
2004	0.59	0.41
2008	0.58	0.42
2012	0.58	0.42
2016	0.60	0.40

peaked as a proportion of the total variation at 0.71 in the 1844 contest between James K. Polk and Henry Clay. It reached its lowest point in 1924, when it contributed only 0.34 of the total variation. Importantly, this was during the height of the Jim Crow era when state-level restrictive voting laws sought to ensure white Democratic dominance in the South. Thus, we shouldn't be surprised that between-state variation was particularly influential in this election and, indeed, that it generally outpaced within-state variation throughout the Jim Crow era. Only as the country moved out of this era and state-based voting impediments began to be reduced (and Democratic dominance of the South began to wane) did within-state variation begin to increase in its effects. Importantly, despite the recent focus on between-state variation in partisan voting, we are actually now close to being back to early 1800s-levels of within-state influence.

The levels of the variation within states also demonstrate significant sub-state variation in the margin variable. The within-state standard deviation ranges from 0.15 in 1972 to 0.34 in 1836. The mean within-state standard deviation from 1828 to 2016 is 0.23.

2.4.1 Spatial Analysis

Given the discovery of this extensive variation in county-level partisan voting, the next step is to diagnose whether there is any spatial patterning to this voting. In order to do so, we employed the global Moran's I test for each presidential election from 1828 through 2016. The results of these global Moran's tests are presented in Table 2.2. As can be seen, there is a strong spatial structuring at the global level in each presidential election. In fact, in all elections, the global Moran is statistically significant at a $p = 0.001$ level (the smallest p-value possible with a permutation approach involving 999 permutations.)

In each election, the spatial autocorrelation is positive. Counties exhibit similar partisan voting as their neighboring counties in each presidential election. The Moran's I values range from 0.488 in 1844 to 0.885 in 1924. The Moran's values, always high, reach their apex in the early and mid-twentieth century before receding in recent decades to the types of values common in the nineteenth century. The principal finding of the global Moran's tests is, however, one of a strong positive spatial patterning in partisan voting at the global level that is statistically significant in all presidential elections.

Table 2.2 Global Moran's I's: 1828–2016

Election	Moran's I
1828	0.725
1832	0.771
1836	0.517
1840	0.546
1844	0.488
1848	0.602
1852	0.599
1856	0.844
1860	0.735
1864	0.604
1868	0.618
1872	0.641
1876	0.685
1880	0.705
1884	0.672
1888	0.710
1892	0.802
1896	0.759
1900	0.760
1904	0.871
1908	0.815
1912	0.828
1916	0.822
1920	0.865
1924	0.885
1928	0.746
1932	0.849
1936	0.831
1940	0.851
1944	0.835
1948	0.742
1952	0.770
1956	0.762
1960	0.679
1964	0.775
1968	0.617
1972	0.582
1976	0.677
1980	0.702
1984	0.535

(continued)

Table 2.2 (continued)

Election	Moran's I
1988	0.501
1992	0.532
1996	0.580
2000	0.585
2004	0.579
2008	0.596
2012	0.615
2016	0.597

The next step is to disaggregate the global results and zoom in at the county level to identify which counties are producing the global patterns and what the spatial patterning of partisan voting looks like in presidential elections. In order to do the latter, we employ a valuable auxiliary tool, the Moran scatterplot. Such a plot displays the observed values on the variable and the weighted average of their neighbor's values as standardized values reflecting how many standard deviations they are above or below the mean. The units' values are plotted on the x-axis and the weighted averages of their neighbors' values are plotted on the y-axis.

The produces a plot in four quadrants. An example of a Moran scatterplot, for the margin variable in the 2016 election, is presented in Fig. 2.2. Significant local Morans in the upper right quadrant reflect positive spatial dependence at values above the mean on the variable. Significant local Morans in the lower left quadrant reflect positive spatial autocorrelation at values below the mean on the variable. Significant local Morans in the bottom right quadrant reflect negative spatial dependence in which units have higher values than their neighbors do. Significant local Morans in the upper left quadrant reflect negative spatial autocorrelation in which units have lower values than their neighbors do.

We can pair the quadrant information from the Moran scatterplot with the significance tests from the local Moran's I's to map the spatial structure of partisan voting in all presidential elections from 1828 to the present. We identify regimes of Democratic and Republican support as well as counties that are spatially uncorrelated with their neighboring counties in their partisan voting. The maps of the spatial regimes in partisan voting for these elections are presented in Figs. 2.3 through 2.50. The counties plotted in blue are counties that are spatially autocorrelated with their neighboring counties in voting that is more Democratic than the national average. Counties plotted in red are counties that are spatially autocorrelated with their neighbors in being more Republican (National Republican in 1828 and 1832 and Whig from 1836–1852) than the national average. The few counties in very dark blue are negatively spatially autocorrelated with their neighbors and are more Democratic than the national average while their neighbors are more Republican than the national average. The few counties in salmon are negatively spatially autocorrelated and are more Republican than the national average while their neighbors are more Democratic than the national average. The counties in white are spatially uncorrelated with their neighboring counties.

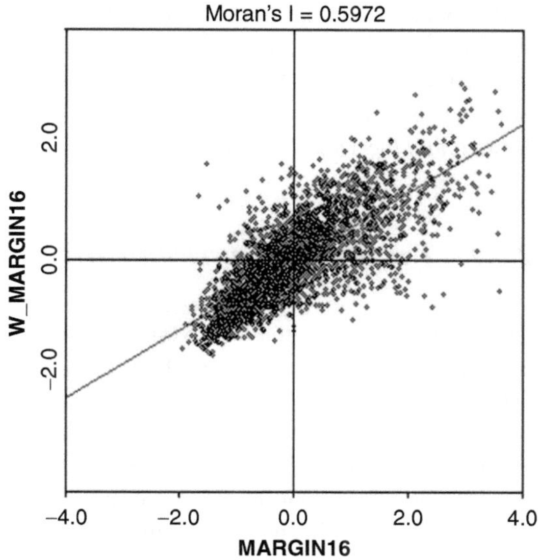

Fig. 2.2 2016 Moran scatterplot

The first, and perhaps most visible, item that becomes clear from the maps is the strong local patterning of partisan voting across American mass voting history. In all of these elections, there has been extensive sub-state and inter-state patterning of partisan voting at the local level. Partisan voting does not simply follow state lines, as a brute interpretation of blue state-red state maps might suggest it does.

Another insight that becomes clear from the maps is that the political geography of partisan voting has changed gradually and in understandable ways over the course of American history. It is generally not marked by dramatic election-specific discontinuities. The pattern instead is one of partisan regimes that ebb and flow in a gradual manner over American history.

2.5 Political Geography and the American Party Systems

We structure our discussion of the political geography of voting in American electoral history around the concept of party systems. The concept of American party systems was developed initially in an influential 1967 book edited by Chambers and Burnham, *American Party Systems: Stages of Political Development*. In this conceptual framework, American political development is divided into a series of sequential party systems. Each party system is defined by a distinct set of issues and policy positions by the two principal American political parties. The First American Party System is generally viewed as running from approximately 1792 through 1824. It featured competition between the rural, limited government Democratic-Republican Party led by Jefferson and Madison and the Federalist Party led by Adams and Hamilton, which had its strongest base of support in the Northeast and supported internal improvements to promote American commerce. This First American Party System predates the period of mass voter participation that was ushered in with the 1828 election. As a consequence, we pick up our analysis of the political geography of mass voting with the beginning of the Second American Party System and the 1828 contest between Andrew Jackson and John Quincy Adams.

2.5.1 The Second American Party System

The Second American Party System covers the period from 1828 through the 1852 presidential election. By 1836, the contours of this party system had been established: vigorous two-party competition throughout much of the United States between the Democratic Party and the Whig Party. The Democratic Party, led by Andrew Jackson (with critical organizational efforts by Martin Van Buren) was intended as a continuation of the old party principles of Jefferson's and Madison's Democratic-Republicans. The Whigs reflected the Federalist Party's commitment to internal improvements financed through a tariff that would also help build the American manufacturing economy, a coherent set of policies known as The American System and promoted by the foremost Whig leader, Henry Clay (Holt 2003).

The Second Party System began in 1828 with a rematch of the 1824 contest between John Quincy Adams (the now-incumbent president) and Andrew Jackson. The 1824 election was one of the most bitter and contested elections in American history. With partisan competition having collapsed during "The Era of Good Feelings", the 1824 presidential election was a competition between four candidates, John Quincy Adams, Andrew Jackson, William Crawford, and Henry Clay, each with regional bases of support. Jackson won a plurality of both the Electoral College and popular votes, but lacking a majority in the Electoral College, he had to compete against Adams and Crawford in the United States House of Representatives' selection of the president. In what Jackson's supporters viewed as a "corrupt bargain", Clay gave his support to Adams, who was chosen by the House members, and was subsequently chosen himself as Adams' Secretary of State.[3]

The contested election of 1824 set up a bitter rematch in 1828. Having been, from his perspective, cheated out of an election that was rightfully his in 1824, Jackson ran a spirited and at times bitter campaign against the man he believed had stolen the election from him. Portraying Adams, a candidate with a long history of governmental experience, as the exemplar of a corrupt Washington establishment, Jackson ran as an outsider candidate whose principal appeal was to the forgotten men of America who formed his base. For their part, establishment politicians such as Adams and Clay were concerned about Jackson's temperament for the presidency and by what they viewed as his lack of support for republican norms (Schlesinger

[3]As Jenkins and Sala (1998) demonstrate, however, there is little support for the "corrupt bargain" thesis. Clay's policy positions were closer to Adams' than they were to Jackson's and the voting in the House is consistent with a model of sincere voting by members for the most like-minded candidate rather than strategic voting as would be consistent with the corrupt bargain thesis.

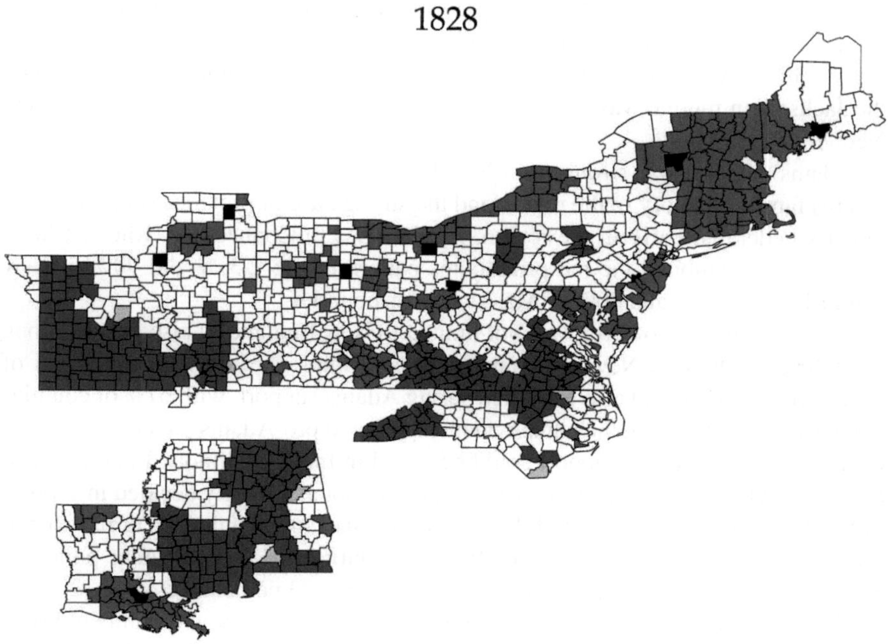

Fig. 2.3 The political geography of county-level voting in the 1828 presidential election

1945). Needless to say, the 1828 presidential election presents clear echoes to the 2016 election, a topic to which we return in Chapter Five.

Andrew Jackson won the 1828 presidential election in an Electoral College landslide, with 178 electoral votes to President John Quincy Adams's 83.[4] The map of the political geography of the 1828 election is presented in Fig. 2.3. In 1828, 22.7% of counties were positively autocorrelated with their neighbors at rates above the national advantage for Jackson. These counties had a mean victory margin of 76 percentage points for Jackson over Adams, indicating particularly strong Jacksonian support levels in these counties.[5] Areas of Jacksonian strength were relatively more common in the Southern and Border states than in the North. In all, spatial regimes of Jacksonian support were found in nine states[6] in 1828, with Pennsylvania counties being the only fully Northern ones to exhibit this pattern.[7]

[4]Electoral College and election results are from Dave Leip's (2018) Atlas of U.S. Presidential Elections, https://uselectionatlas.org/

[5]Because our interest is in the political geography of county-level voting and in counties as units of analysis, the averages reported in this chapter are not weighted by population.

[6]Virginia counties in what would later be West Virginia also had strong Jacksonian support levels, but West Virginia would not become a state until 1863.

[7]The Illinois counties exhibiting strong Jackson support were located in the Southern portion of the state. This regime extended westward into Missouri, a border state.

Alabama, Missouri, and Mississippi were particularly states that included Jacksonian fervor. 67% of Alabama counties evidenced the blue regime in Fig. 2.3 while 63% of Missouri counties and 45% of Mississippi counties did. Conversely, the Jacksonian regime was completely absent in states such as Ohio, Indiana, and New York.

Adams' support was strongest in New England. This is not surprising given the Adams family's history in the region and the strong base of previous support for his father's Federalist Party in the region. All counties in Connecticut, Rhode Island, and Vermont exhibited the red pro-Adams pattern in Fig. 2.3, as did 90% of New Hampshire counties and 86% of Massachusetts counties.

Adams' support, however, was not limited to New England. It also existed along Lake Erie in Western New York and Northern Ohio. The mid-Atlantic states of Delaware and Maryland also exhibited strong Adams support, with 67% of counties in the former and 64% in the latter evidencing the red pro-Adams pattern in Fig. 2.3. Other pockets of Adams support could be found in Indiana, Illinois, Kentucky, and Louisiana. Overall, the pro-Adams counties were not as strongly decided in support of the National Republican candidate as the pro-Jackson counties were in favor of Jackson. Where Jackson's margin in the blue counties was 76 percentage points on average, Adams' margin in the red counties was only 23 percentage points.

The 1832 presidential election featured a contest between now-incumbent President Andrew Jackson and National Republican candidate Henry Clay, John Quincy Adams's Secretary of State. Jackson easily won re-election, with an expanded Electoral College margin of 219 to 49 over Clay (with a scattering of electoral votes also cast for minor party candidates John Floyd and William Wirt). Jackson also won 55% of the popular vote to Clay's 37%.

The map of the political geography of the 1832 presidential election is presented in Fig. 2.4. The 1832 map is quite similar to the 1828 map. In the Jackson vs. Clay contest, Jackson's support was again strongest in Missouri and in the South, with Alabama and Mississippi again bastions of Jacksonian (now Democratic Party) strength. In the North, counties in central Pennsylvania and southern Illinois were again strongly pro-Jackson. In all, blue regimes of Democratic support were found in nine states (Alabama, Illinois, Louisiana, Missouri, Mississippi, North Carolina, Pennsylvania, Tennessee, and Virginia).

Clay's areas of support in the Northeast also replicated Adam's areas from four years before. Clay had a red regime of strong support in Connecticut, Massachusetts, Maine, New Hampshire, New York, Rhode Island, and Vermont. To this Clay also added an expanded regime of support in Ohio and Indiana as well as support in his home state of Kentucky. However, where Jackson ran strongly, he ran very strongly, much more so than for Clay in his own redoubts of support. Jackson's average margin of victory in the blue counties in 1832 was an amazing 91.6 points, the largest average margin in absolute terms for any presidential candidate in history. In contrast, Clay enjoyed an average margin of only 7.6 points in his red counties. And in the spatially uncorrelated counties in 1832, Jackson enjoyed an average margin of victory of 41.5 points.

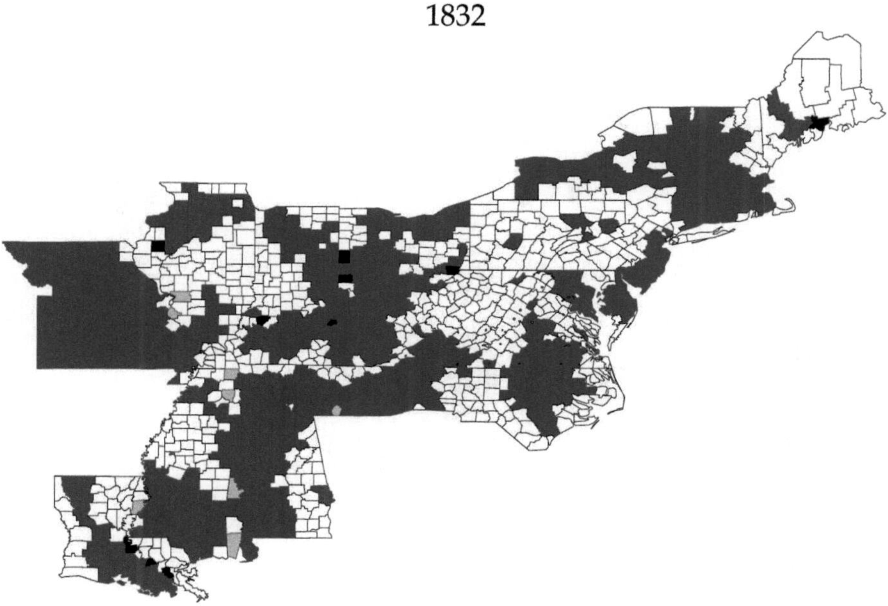

Fig. 2.4 The political geography of county-level voting in the 1832 presidential election

By 1836, the competition between the Democratic and Whig parties that would mark the rest of the Second American Party System was set. In this election, the Democratic candidate, Martin Van Buren, defeated a set of regional Whig candidates (William Henry Harrison, Hugh Lawson White, and Daniel Webster) who competed in regions of expected strength in an effort to deny Van Buren an Electoral College majority.[8]

The map of the 1836 election in Fig. 2.5 highlights the full emergence of one of the defining characteristics of the Second American Party System: vigorous two-party competition throughout the country. No longer do we see the areas of regional strength that were present in 1828 and 1832. Instead, we see pockets of blue and red throughout the country. This, of course, is clearly a reflection of the failure of the Whigs' regional strategy in this election. Had their strategy worked, we should have seen the regional Whig candidates dominate Van Buren in each of the three areas in which they competed. Instead, Van Buren won a clear Electoral College victory and became the last Vice President to succeed a two-term President of his own party until George H.W. Bush in 1988. But perhaps reflecting his lack of Jackson's charisma, Van Buren's county-level margins of victory were down considerably

[8]Harrison was the Whig candidate on the ballot in CT, DE, IL, IN, KY, ME, MD, MI, NH, NJ, NY, OH, PA, RI, and VT. White was the Whig candidate on the ballot in AL, AR, GA, LA, MS, MO, NC, TN, and VA. Webster was the Whig candidate on the ballot in MA. The results presented in Fig. 2.4 are for the contest between Jackson and the specific Whig candidate in each state.

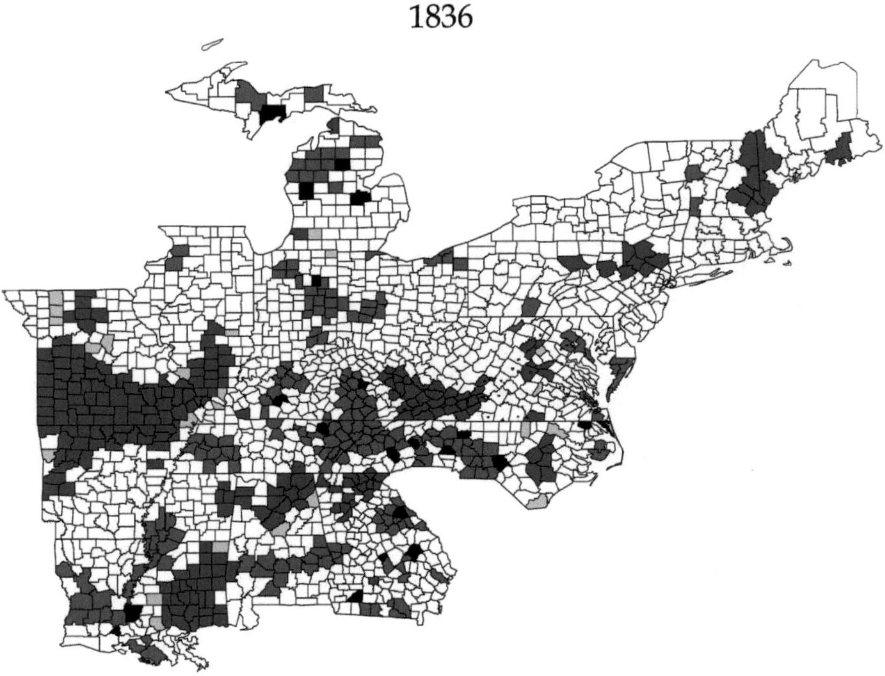

Fig. 2.5 The political geography of county-level voting in the 1836 presidential election

from the Jackson victories in 1828 and 1832. Conversely, Whig margins were up considerably from Clay's National Republican candidacy in 1832, foreshadowing Harrison's victory in the 1840 election.

The geographically diffuse nature of partisan voting in 1836 can be seen in the fact that regimes of Democratic support could be found in 17 states while regimes of Whig support could be found in 15 states. Democratic support could be found in states as geographically distant as Arkansas and New Hampshire while Whig support existed in states as distant as Louisiana and New York. One of the clear takeaway points from the 1836 map is the lack of state-level structuring. Partisan regimes existed as pockets of partisan strength within and across state boundaries.

The 1840 presidential election featured a rematch between the two leading opponents from the 1836 election, Democrat Martin Van Buren and, now running as the sole Whig candidate, William Henry Harrison. Following the Panic of 1837 and ensuing recession that took place on Van Buren's watch, Harrison easily bested Van Buren in the Electoral College 234–60, although the popular vote was closer, with only a six point margin separating the two candidates.

The 1840 map in Fig. 2.6 reflects the maturing of the nation's new two-party system. Many of the areas of Democratic and Whig support in 1836 repeated themselves in 1840. The spatial regime of strong Democratic support in northern Arkansas, southern Missouri, and southern Illinois replicated in 1840, as did the

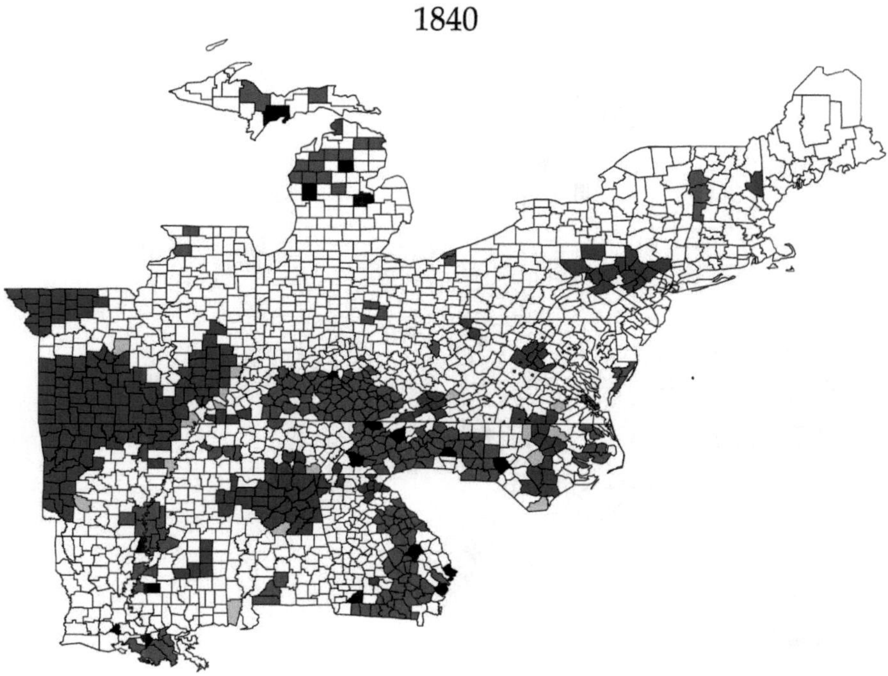

Fig. 2.6 The political geography of county-level voting in the 1840 presidential election

pockets in northern Alabama and Eastern Pennsylvania. Meanwhile, the Whig support in upstate Michigan and the upper peninsula of Michigan repeated in 1840 as did the regime of Whig support in eastern Tennessee and western North Carolina.

Further reflecting the maturation of the two-party system, for the first time Democratic support in its strongholds lagged behind support for the opposing party. The average victory margin in the red counties in Fig. 2.6 was 47.9% while the average victory margin in the blue counties was only 41%.

The 1844 contest featured James K. Polk, a Jacksonian Democrat, against Whig candidate Henry Clay. The central issue of the election was the possible annexation of Texas, a move favored by Polk and opposed by Clay unless the North agreed to this annexation. Very importantly, despite the introduction of a new issue intimately connected both with the concept of Manifest Destiny and the increased salience of slavery (due to its spread with the potential inclusion of Texas in the Union) (Bicknell 2014), the map of 1844 was little changed from those in 1836 and 1840.

As Fig. 2.7 shows, we see the same basic patterns in 1844 that we saw in the preceding two elections. The same area of strong Democratic support in northern Arkansas, southern Missouri, and southern Illinois repeated itself again as did the Democratic pattern in upstate Alabama. So did the areas of Whig support in upstate Michigan, Kentucky, eastern Tennessee, and western North Carolina.

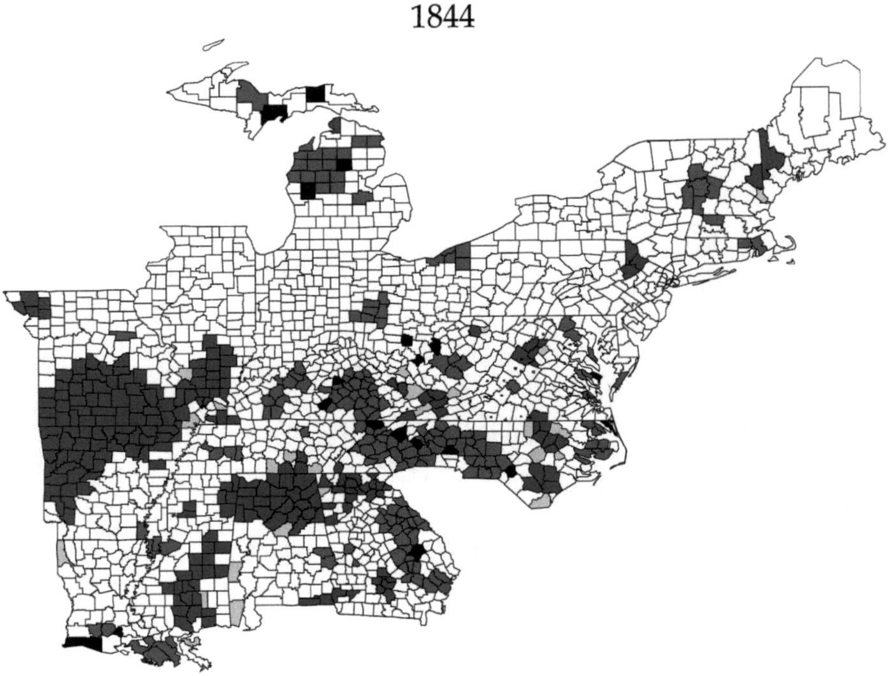

Fig. 2.7 The political geography of county-level voting in the 1844 presidential election

The institutionalization of the Democratic-Whig competition in the Second American Party System from 1836 through 1844 can be seen by comparing the states that evidenced Democratic and Republican spatial regimes in these elections. In all, blue Democratic regimes could be found in 17 states in 1836: Alabama, Arkansas, Georgia, Illinois, Kentucky, Louisiana, Maine, Michigan, Missouri, Mississippi, North Carolina, New Hampshire, New Jersey, New York, Pennsylvania, Tennessee, and Virginia. The pattern would exist in 15 states in 1840, with all but two of the 1836 states (Louisiana and Maine) repeating the pattern in 1840 and no new states joining it. In 1844, 15 states would again exhibit the pattern, with only New Jersey and New York absent from the 1836 group and no new states that hadn't exhibited this Democratic pattern joining the group in 1844.

Turning to the Whig side of the ledger, 15 states had regimes of Whig support in 1836: Alabama, Georgia, Illinois, Indiana, Kentucky, Louisiana, Maryland, Michigan, Mississippi, North Carolina, Ohio, Pennsylvania, Tennessee, Virginia, and Vermont. The same basic pattern would present itself again in 1840, with only Arkansas joining this group of states and Pennsylvania leaving it. Slightly more variation did appear in 1844, perhaps due to the emerging concerns about slavery and the annexation of Texas. Massachusetts joined the group in this election, along with Missouri, New York, and Rhode Island, while Illinois left the 1840 group of states. In all, however, the story of Whig support in 1844 is much more one of continuity from 1840 than change.

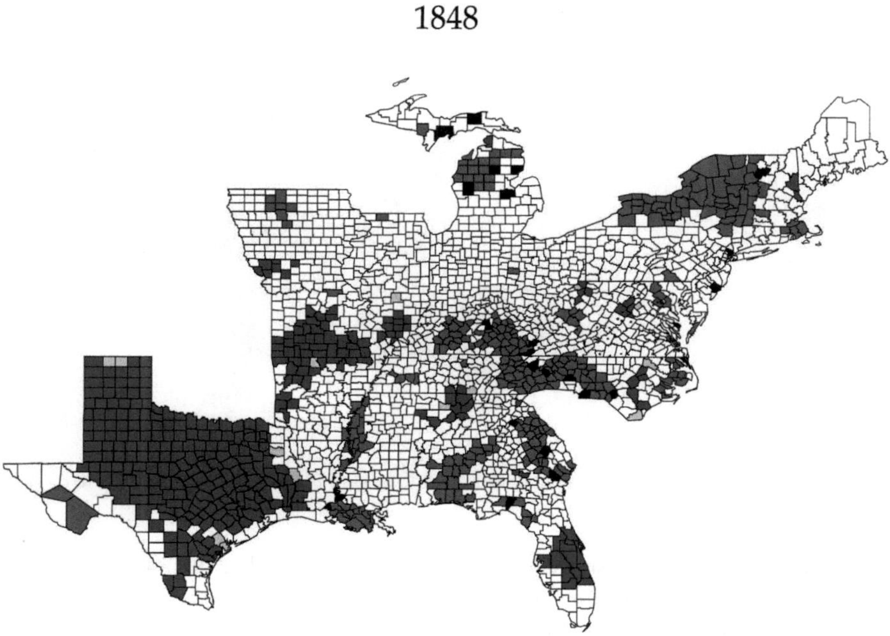

Fig. 2.8 The political geography of county-level voting in the 1848 presidential election

The map of the 1848 election in Fig. 2.8 presents an expanded United States, now including Texas, Florida, and Iowa, added to the Union since the 1844 election. Unsurprisingly, counties in Texas were strongly Democratic in the contest between Democrat Lewis Cass and Whig Zachary Taylor. The average margin for Cass in these Texas counties was 54 percentage points. Newly added Florida likewise evidenced a regime of Democratic support while also evidencing a Republican regime in its panhandle.

Beyond this, however, even with the expanded map of 1848, we see some continuity with the preceding three elections in the Second Party System. Again we see the Democratic pattern in northern Arkansas, southern Missouri, and southern Illinois as well as in northern Alabama. And we see the familiar regimes of Whig support in upstate New York, Kentucky, eastern Tennessee, and western North Carolina. Perhaps the biggest change in the 1848 map, in addition to the Democratic terrain in recently-annexed Texas, is the strong Whig support in New York that also bled over into Vermont. The average margin of victory for Taylor in these New York counties was 25 percentage points while it was 34 points in these Vermont counties.

Taylor's election in 1848 would end up being the last election victory for a Whig presidential candidate. The 1852 election, the last in the Second Party System, featured a contest between Democrat Franklin Pierce and Whig Winfield Scott. Pierce easily bested Old Fuss and Feathers Scott in an Electoral College landslide. And yet despite an election that was far more lopsided in the Electoral College than

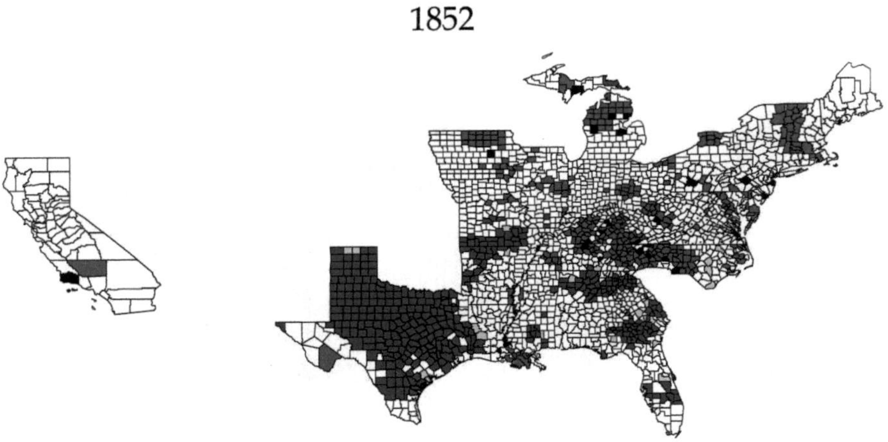

Fig. 2.9 The political geography of county-level voting in the 1852 presidential election

the preceding ones, the basic patterns of Democratic and Whig support were largely unchanged from the preceding elections. We will, in fact, see this later on as well—while means vary between landslide and non-landslide elections, relative areas of Democratic and Republican support in later elections often remain unchanged across elections.

Again we see the pattern in Fig. 2.9 of Democratic support in Texas, northern Arkansas and southern Missouri (less so in southern Illinois in this election) and northern Alabama. And we see the familiar patterns of Whig support in upstate Michigan, Kentucky, eastern Tennessee, and western North Carolina. The dominant Whig regime in New York from 1848 was much diminished in 1852, with only 18% of New York counties exhibiting the Whig regime in this election. But the pattern continued from 1848 in Vermont, with 79% of Vermont counties exhibiting the Whig regime. In all, with the end of the Second American Party System, the familiar patterns of voting from 1836–1852 would also disappear following the 1852 election.

2.5.2 The Third American Party System

The Second American Party System, with its competitive two-party system in all regions between the Whigs and the Democrats, fell apart in the 1850s with the emergence of slavery as a dominant political cleavage in the nation. The Whigs, unable to take a strong position in opposition to or in support of slavery, were replaced as the main competition to the Democrats by the anti-slavery Republican Party by the 1856 election. And with the emergence of the Republican Party by 1856, the Third American Party System began, running until the 1890s.

There were several issues that animated the Third American Party System. First and foremost was the Civil War and its aftermath. The Republican Party's identification with the successful prosecution of the war—and the Democratic Party's identification with rebellion—would shape not only the Civil War election of 1864 and Reconstruction, but also lead to a successful waving of the "bloody shirt" for decades after the war as Republicans reminded voters of the Democrats' identification with the Confederacy. The aftermath of the Civil War also shaped politics as the two parties took diverging positions on tariffs that funded Union pensions.

In additional to the sectional issues of the Third Party System, there was also a strong pietist vs. liturgical divide, as religious denomination became strongly correlated with partisan identification (Kleppner 1979). Pietist denominations such as Methodists, Congregationalists, and Presbyterians, became strongly associated with the Republican Party and believed in using the government to enforce moral values and to use public schools to promote Protestant pietist morality. Liturgically-based denominations like Roman Catholics and Episcopalians, believed that personal morality, including on drinking, should not be enforced by the government, and Catholics favored parochial, Catholic schooling in opposition to Protestant-dominated public schooling. The liturgical denominations' positions became closely identified with the Democratic Party as did the voting behavior of these denominations' members.

The 1856 presidential election pitted Democratic candidate James Buchanan (who had been nominated over the sitting Democratic President, Franklin Pierce), Republican John C. Frémont, and American Party candidate (and former Whig President) Millard Fillmore. Fillmore, running on a nativist platform, carried only Maryland. As the initial Republican candidate for President, Frémont fare far better, carrying 11 states with 114 electoral votes in an impressive show of sectional strength in the North. Still, Buchanan was the victor in the 1856 election, carrying 19 states with 174 electoral votes.

From a geographic perspective, perhaps the most striking feature of the 1856 election was the strong sectional performance by Frémont in much of the North. A continuous spatial regime of autocorrelated Republican support above the national average extends in Fig. 2.10 from the easternmost county in Maine to the westernmost county in Iowa. In all, the spatial regime in the North stretched approximately 1500 miles, reflecting the emerging sectional division in the nation.

In all, this pattern includes every county in Connecticut, Maine, New Hampshire, Rhode Island, and Vermont, as well as 89% of the counties in Iowa, 86% of the counties in Massachusetts, 84% of the counties in New York, 82% of the counties in Ohio and Wisconsin, 77% of the counties in Michigan, 58% of the counties in Indiana, 40% of the counties in Illinois, and 33% of the counties in Pennsylvania. Importantly, however, for both Frémont and the Republicans, while Frémont ran well in these counties with an average margin of 18.6 percentage points, this paled in comparison to Buchanan's advantage in the blue counties in Fig. 2.10 as well as the spatially uncorrelated counties in white.

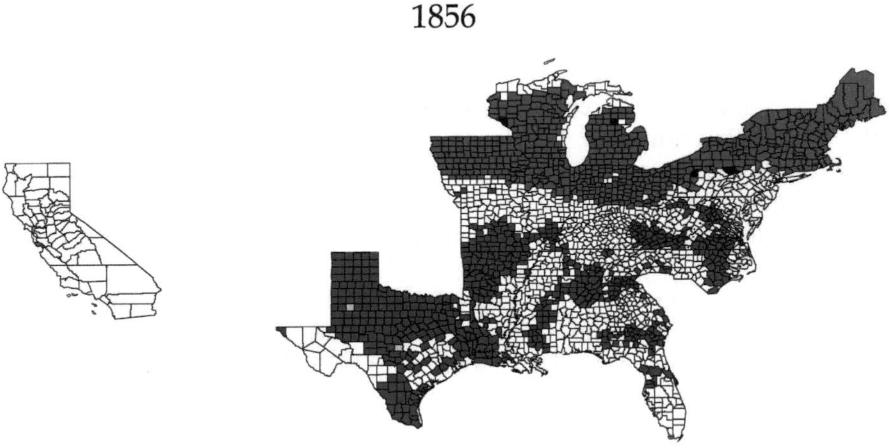

Fig. 2.10 The political geography of county-level voting in the 1856 presidential election

Buchanan's average margin of victory in the blue counties was 71.6 percentage points. In the spatially random counties, which comprised 50% of all counties in 1856, Buchanan's margin was 48 percentage points. The problem for Frémont, as the first Republican candidate, was that while he ran reasonably well where he ran well, he didn't run well enough there, or in enough places, to win the presidency. This would have to wait four years for the second Republican nominee for President, Abraham Lincoln.

Buchanan's presidency was marked by its inability to address the growing divisions in the nation regarding slavery. Buchanan did not pursue a second term in office in 1860 and was succeeded instead as Democratic nominee by Senator Stephen A. Douglas of Illinois. The 1860 election would, in fact, feature a four-way contest. Douglas's principal opponent was Abraham Lincoln, the Republican nominee and his opponent in the 1858 Senate contest in Illinois (and in the famous Lincoln-Douglas debates). The major party nominees Douglas and Lincoln were joined in the 1860 presidential contest by Buchanan's Vice President John C. Breckinridge, the nominee of the Southern Democrats, and Tennessean John Bell, the nominee of the Constitutional Union Party.

Lincoln would win in an Electoral College landslide, with 180 Electoral College votes to Breckinridge's 72, Bell's 39, and Douglas's 12. In the popular vote, Lincoln would best Douglas by more than 10 percentage points, 39.7% to 29.5%. Shortly after Lincoln's election, the Southern states seceded from the Union, and the Civil War ensued.

The political geography of the 1860 presidential election is presented in Fig. 2.11. As we can see, Douglas fared well in counties in Missouri (the only state he won in the Electoral College) as well as in the Southern portion of his shared home state with Lincoln, Illinois, as well as in portions of Kentucky, Texas, Alabama, Georgia, and other states. Douglas's average margin over Lincoln in the blue

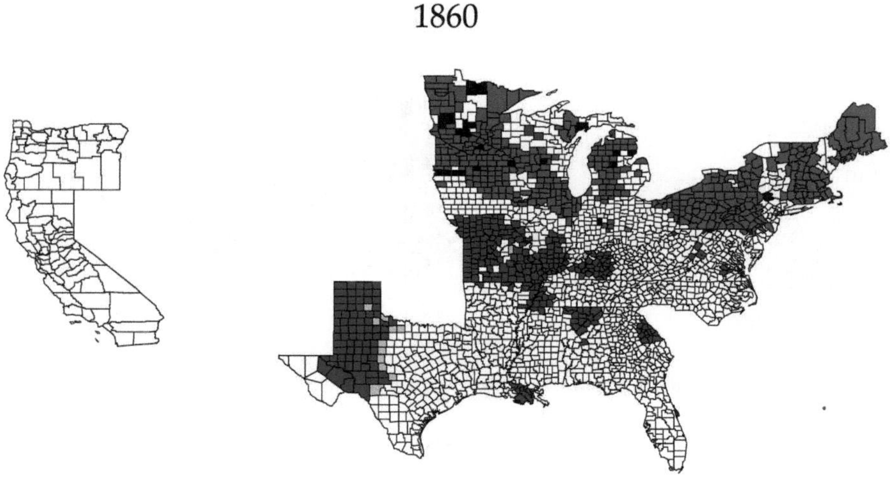

Fig. 2.11 The political geography of county-level voting in the 1860 presidential election

counties in Fig. 2.11 was 39.1 percentage points. In contrast, Lincoln fared well in New England, Pennsylvania, New York, Michigan, Wisconsin, Minnesota, and in the northern portions of Illinois. Lincoln's average margin over Douglas in these red counties was 39 percentage points.

The 1864 presidential election featured a contest between the incumbent President, Abraham Lincoln (now running as the candidate of the National Union Party that brought together Republicans and War Democrats) against Democratic candidate (and former Commanding General of the United States Army) George B. McClellan. The Civil War, of course, was the defining issue of the contest. McClellan headed up a divided Democratic Party, reflected by the balancing of his pro-war stance with the anti-war stance of his running mate, George Pendleton. Buoyed by multiple military victories, Lincoln won an easy re-election victory in the Electoral College, carrying 221 electoral votes to McClellan's 21.

McClellan's three electoral college victories in Delaware, Kentucky, and New Jersey are reflected in the blue spatial patterns in Fig. 2.12. McClellan also did well in parts of Wisconsin (in the Northern part of the state and in Eastern counties along Lake Michigan), in Southern Illinois, and in parts of Maryland. Lincoln did particularly well in Vermont, western Missouri, Kansas, and portions of Iowa and Minnesota. In all, his margin of victory in the red counties in Fig. 2.12 averaged 65 percentage points. He also enjoyed an average margin of 14 percentage points in the spatially random counties in Fig. 2.12, reflecting his geographically broad victory in this election.

The first post-Civil War election occurred in 1868, during the Reconstruction Era. The contest featured Republican (and military victor in the Civil War as Commanding General of the Army) Ulysses S. Grant against Democrat (and former New York Governor) Horatio Seymour. The principal issue of the election was how Recon-

1864

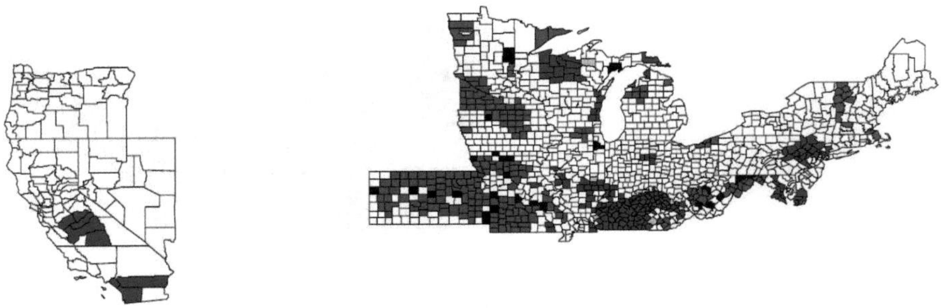

Fig. 2.12 The political geography of county-level voting in the 1864 presidential election

1868

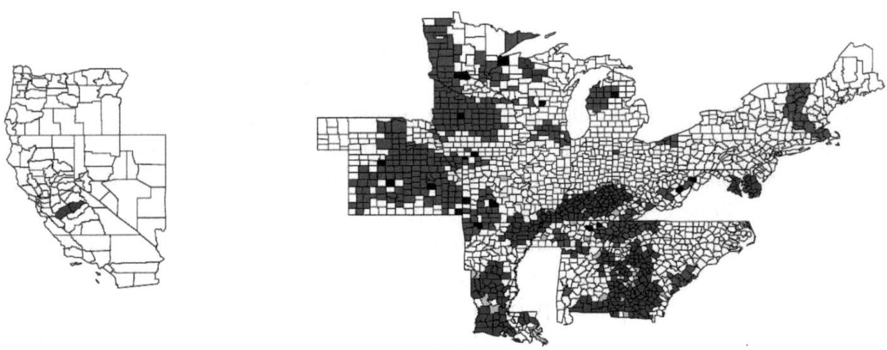

Fig. 2.13 The political geography of county-level voting in the 1868 presidential election

struction should be conducted, with Grant supporting congressional opposition to impeached incumbent President Andrew Johnson's Reconstruction policies and Seymour favoring a conciliatory approach toward the South (Foner and Mahoney 1995). Although Grant won a substantial Electoral College victory, the popular vote was relatively close, and surprisingly so in the immediate wake of a Civil War in which the two parties were so closely identified with the competing sides.

As we can see from Fig. 2.13, several Southern states did not take part in this Reconstruction Era election while one Southern state, South Carolina, experienced popular voting for President for the first time. The low country of South Carolina was solid Republican territory while its upstate was solid Democratic territory, much like most of neighboring Georgia. Spatial regimes of Democratic support could also be found in Southern states such as Alabama, Arkansas, and Louisiana and in Border states such as Kentucky, Maryland, and Tennessee, as well as in northern states such as Delaware and Wisconsin.

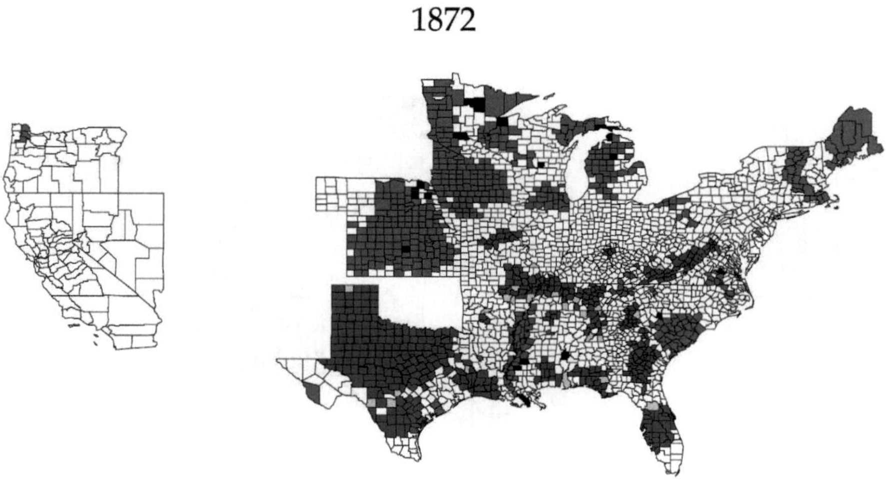

Fig. 2.14 The political geography of county-level voting in the 1872 presidential election

Grant's support base in 1868 consolidated Lincoln's from 1864 while producing spatial regimes that would become common throughout much of the rest of the Third American Party System. Spatial regimes of red Republican support could be found in Kansas, Minnesota, southern Wisconsin, western Michigan, and Vermont. These patterns would continue for decades.

Interestingly, Seymour's margins in the blue counties were, on average, larger than the margins for Grant in the red counties. On average, Seymour enjoyed a 58.2 percentage point margin in the blue counties while Grant enjoyed a 52.9 percentage point margin in the red counties. The spatially uncorrelated counties tilted slightly in Grant's favor, with an average Republican margin of 6.4 percentage points.

Still, one of the principal implications of the 1868 election was the reestablishment of the Democratic Party as a credible national competitor to the Republicans. The relative closeness of the presidential contest would be reflected in many subsequent Third Party System presidential elections.

The 1872 election was a bit anomalous, featuring incumbent President Ulysses S. Grant against newspaper editor and Liberal Republican Party nominee Horace Greeley. Greeley's candidacy reflected a division within the Republican Party and he was subsequently endorsed by the Democratic Party as their standard bearer as well. Despite the unique characteristics of the 1872 election, several of the features of the political geography of voting that began in 1868 repeated themselves in 1872. Once again, in Fig. 2.14, we can see the regimes of Republican support in Kansas, Minnesota, southern Wisconsin, western Michigan, and Vermont. Regimes of support for Grant could also be found in Arkansas, Connecticut, Illinois, Iowa, Kentucky, Louisiana, Massachusetts, Maine, Mississippi, North Carolina, Nebraska, New Hampshire, New Jersey, New York, Ohio, Oregon, Pennsylvania, Rhode Island, South Carolina, Tennessee, and Virginia. On average, Grant's margin in these counties was 49.6 percentage points.

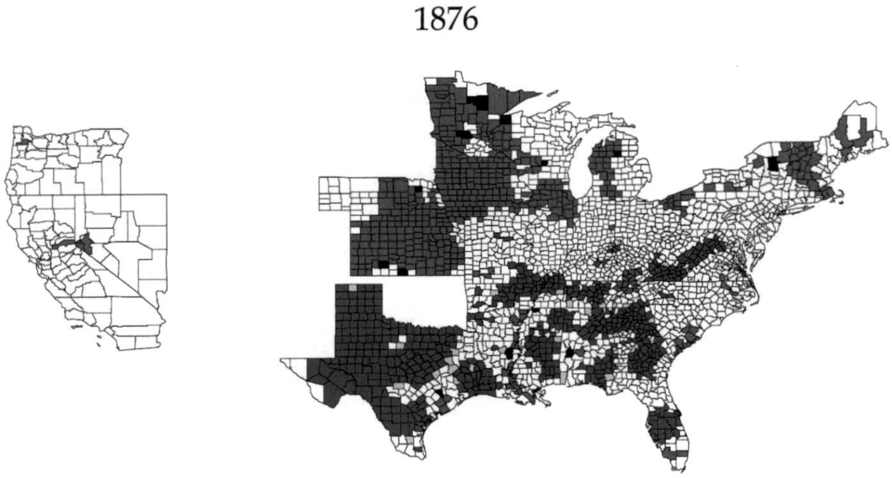

Fig. 2.15 The political geography of county-level voting in the 1876 presidential election

Greeley, however, was equally strong in the blue counties in Fig. 2.14, where he enjoyed an average 49.8 percentage point margin of victory. Greeley ran strong in Texas, where 73% of counties exhibited the blue spatial regime. Regimes of Greeley support were also found in Alabama, Arkansas, Florida, Georgia, Illinois, Kentucky, Louisiana, Missouri, Mississippi, North Carolina, Tennessee, Virginia, and West Virginia. In all, 18.4% of counties exhibited the red Republican spatial regime while 16.7% exhibited the blue Liberal Republican/Democratic spatial regime. In the 63.1% of counties that were not spatially autocorrelated, Grant enjoyed a 5.8 percentage point margin of victory on average.

The 1876 election was one of the more controversial elections in U.S. history, matched only by the 2000 election in its contested nature and the involvement of the U.S. Supreme Court. Democratic candidate and New York Governor Samuel Tilden faced off against Republican candidate and Ohio Governor Rutherford B. Hayes. Tilden won a clear majority in the popular vote and was ahead by 19 votes in the Electoral College, with electoral college votes from four states (Florida, Louisiana, Oregon, and South Carolina) contested. Eventually through the Compromise of 1877, the 20 electoral college votes from these states were awarded to Hayes in exchange for the removal of federal troops from the South, the end of Reconstruction, and the eventual ceding of the South to the Democrats for decades (Holt 2008).

The 1872 and 1876 elections were, in short, quite different elections. The former took place with a popular incumbent running for reelection against a Liberal Republican candidate endorsed by the Democrats. The latter took place after two terms of corruption in the Grant administration with a relatively unknown Ohio governor facing off against a prominent reformist Democratic governor. And yet, the map of spatial regimes in 1876 in Fig. 2.15 is quite similar to the 1872 map in Fig. 2.14.

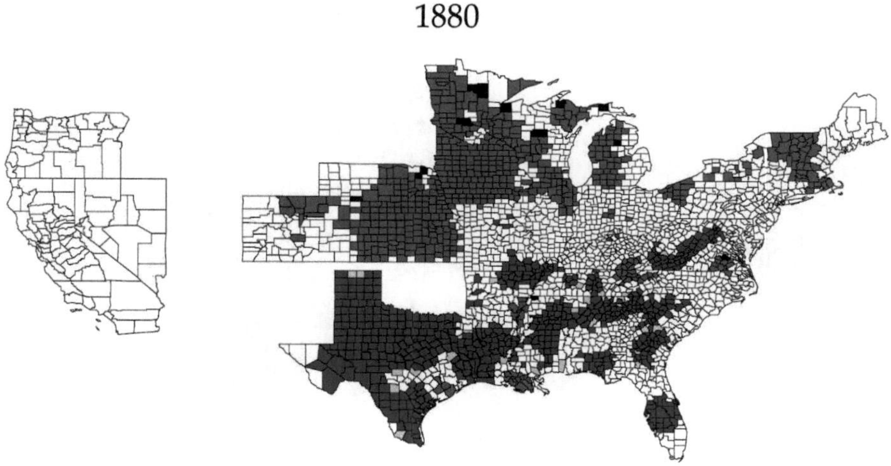

Fig. 2.16 The political geography of county-level voting in the 1880 presidential election

The Democratic spatial regimes in states such as Florida, Tennessee, Texas, Virginia, and West Virginia in 1872 were replicated in 1876. Likewise, Republican spatial regimes from 1872 in states like Kansas, Michigan, Minnesota, Nebraska, and Vermont repeated themselves in 1876. In short, the map in Fig. 2.15 indicates that party bases in the Third American Party System were becoming solidified.

The 1880 presidential election featured a faceoff between two former Union officers from the Civil War. The Republican candidate was former Major General (and former Representative from Ohio) James Garfield while the Democratic candidate was Union General (and one of the heroes of the Battle of Gettysburg) Winfield Scott Hancock from Pennsylvania. This was the first presidential election conducted after Reconstruction and after the removal of federal troops from the South. And although this led, as expected, to Hancock carrying the South (with Garfield carrying nearly all states in the North), the *relative* areas of Democratic and Republican support in 1880 were largely unchanged from the less sectional (in the Electoral College) election of 1876, as can be seen in Fig. 2.16.

In all, spatial regimes of Republican support could be found in 26 states in 1880. And regimes of Democratic support could be found in 14 states. Garfield's average margin of victory in the red counties in Fig. 2.16 was 31.8 percentage points. The support for Hancock was in fact, much stronger in the blue counties in Fig. 2.16, averaging 61.8 percentage points, reflecting the solidifying of solidly Democratic Southern counties. Hancock also enjoyed an average 8.1 percentage point margin in the spatially uncorrelated counties in Fig. 2.16. Garfield was advantaged, however, in carrying the more populous Northern states, which carried greater weight than the less populous Southern states in the Electoral College.

The 1884 election featured a contest between Democratic nominee and New York Governor Grover Cleveland and Republican nominee and former Senator James G. Blaine. Blaine, seriously considered for previous Republican nominations, had been

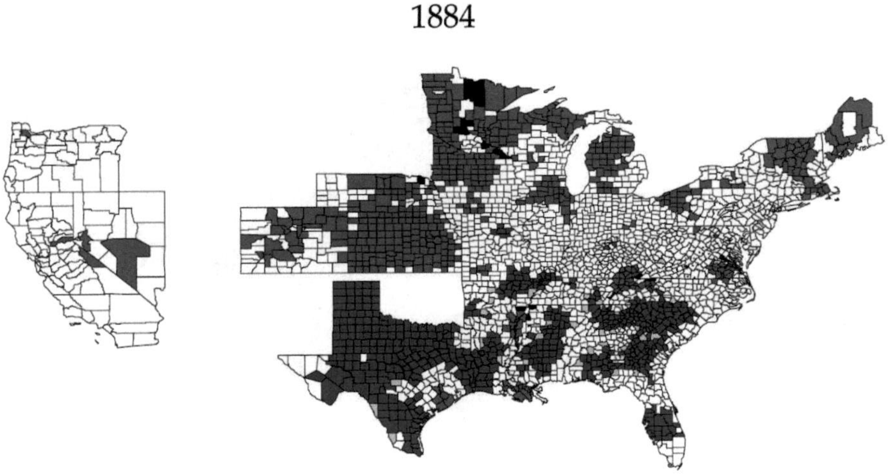

Fig. 2.17 The political geography of county-level voting in the 1884 presidential election

dogged by concerns about public corruption for years. Cleveland, conversely, had earned a reputation as a politician opposed to corruption and a political reformer who, among other things, took on Tammany Hall, the Democratic machine in New York City. Cleveland was aided by one of the first "October surprises" when, just a few days before election day, Reverend Samuel Burchard referred to the Democratic Party as the party of "rum, Romanism, and rebellion" (Summers 2000). Cleveland won the election narrowly in the Electoral College, becoming the first Democratic nominee to win the presidency since James Buchanan in 1856.

Yet for all the unique features that make the 1884 contest so intriguing, the spatial regimes in this election largely reflected the patterns that had become established by now in the Third American Party System, as can be seen in Fig. 2.17. Cleveland ran strong in the Southern states in which previous Democratic candidates had run well. Blaine ran well in the areas of the North where previous Republican candidates had run strong. This suggests the clear maturation of a party system in which the two parties had developed localized bases of partisan support.

In all, 18.5% of counties in 1884 reflected the blue spatial patterns in Fig. 2.17, while 20.1% reflected the red spatial regimes. Where Cleveland ran strongly, he ran strong, indeed much more strongly than Blaine did in his bastions of Republican support. Cleveland's average margin of victory in the blue Democratic regimes was 61.8 percentage points while Blaine's average margin of victory in the red Republican redoubts was only 25.7 percentage points. In the spatially uncorrelated counties in Fig. 2.17, Cleveland ran ahead of Blaine by, on average, 4.9 points.

The 1888 presidential election pitted incumbent Democratic President Grover Cleveland against former Indiana Republican Senator Benjamin Harrison (grandson of former Whig President William Henry Harrison). Cleveland and Harrison were attractive as nominees to their two respective parties because New York and Indiana

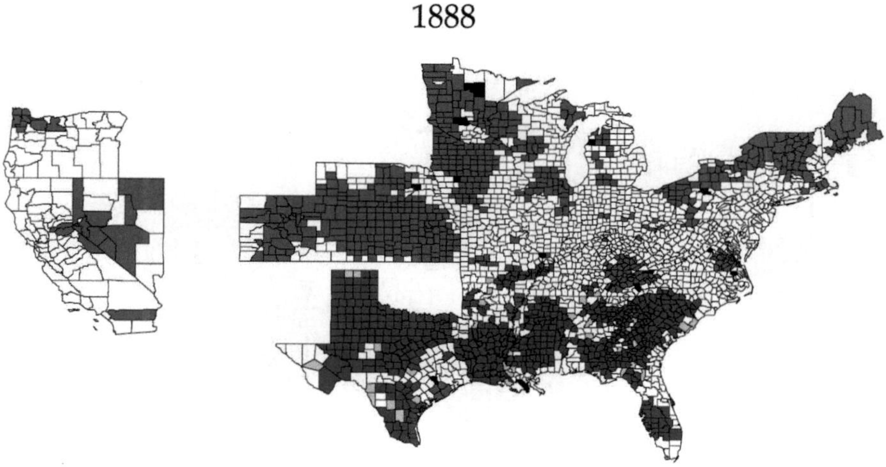

Fig. 2.18 The political geography of county-level voting in the 1888 presidential election

were among the leading swing states during the Gilded Age and thus nominating a candidate who hailed from one of these states could help swing the Electoral College. In the end, Cleveland ended up narrowly losing the Electoral College to Harrison despite winning the popular vote.

The 1888 map of spatial regimes, in Fig. 2.18, bears similarities to preceding elections, but also some dissimilarities that would be harbingers of the political geography of elections during the subsequent Fourth American Party System. Again, the Republican nominee fared well in Kansas, Nebraska, Minnesota, Southern Wisconsin, and Western Michigan. And again, the Democratic nominee fared well in Texas, Florida, and Georgia.

Cleveland's Southern strength in 1888, however, was broader than it had been in 1884. The blue spatial regime was starting to spread across the South, reflecting the consolidation of the solid Democratic South following the end of Reconstruction through restrictive voting laws, violence, intimidation, and other means. And Republican support extended farther westward across New York than it had in preceding elections. Likewise, there was a larger area of Republican support in western states such as Nevada than in preceding elections. In short, the strongly sectional Fourth American Party System was being hinted at in the political geography of the 1888 election.

The 1892 election featured a rematch between the candidates from the preceding election, Cleveland and Harrison. This time, Cleveland won both the popular vote and the Electoral College, becoming the first and only president to serve two non-consecutive terms. Cleveland and Harrison were joined in the contest by Populist Party candidate James B. Weaver. Weaver fared quite well in the West, winning Electoral College votes in five Plains and Mountain states. In the end, however, the dominant story was Cleveland's return to office on a policy of strict money.

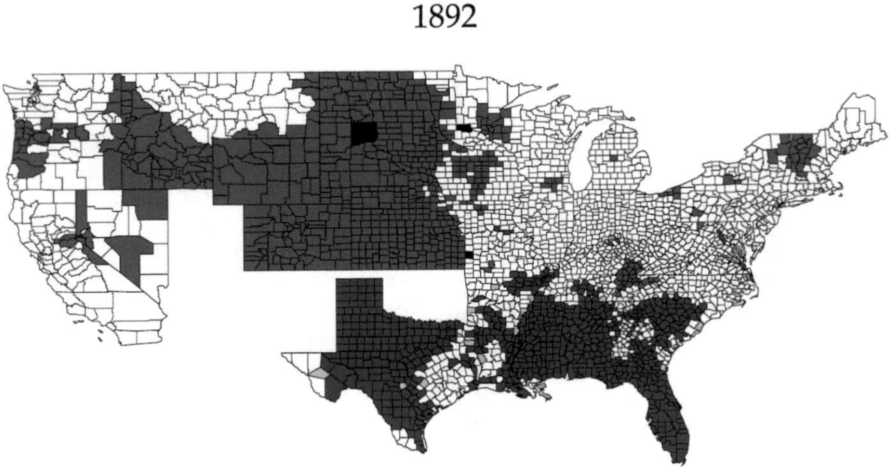

Fig. 2.19 The political geography of county-level voting in the 1892 presidential election

As Fig. 2.19 shows, the South was nearly solidly Democratic in the 1892 election, with a spatial regime that included Texas, Florida, and much of the Deep South. In contrast to 1888, there was not the strong Northern pattern of support for Harrison, in part no doubt due to the lackluster nature of his administration. Instead, the regime of Republican support was located in the Plains and Mountain states. Cleveland enjoyed an average margin of 59.2 percentage points in the blue spatial regimes while Harrison enjoyed an average margin of 36.6 percentage points in the red spatial regimes. The uncorrelated counties went for Cleveland by an average margin of 4.8 percentage points.

2.5.3 The Fourth American Party System

As the passions associated with the Civil War increasingly receded in memory and the Industrial Revolution proceeded apace, the Third American Party System was succeeded in the 1890s by the Fourth Party System. Where waving the bloody shirt and tariff debates dominated much of the Third Party System's discourse, the Fourth Party System was focused primarily on channeling the energies of the Industrial Revolution while ameliorating, to a limited extent, its economic and political dislocations. These ameliorations were promoted via a series of Progressive Era reforms.

The Fourth American Party System saw a decline in national party competition in comparison to the Third Party System. Where Republicans typically had a small edge in close national elections in the Third System, they became the dominant party in the Fourth System, winning all but two presidential elections during this era

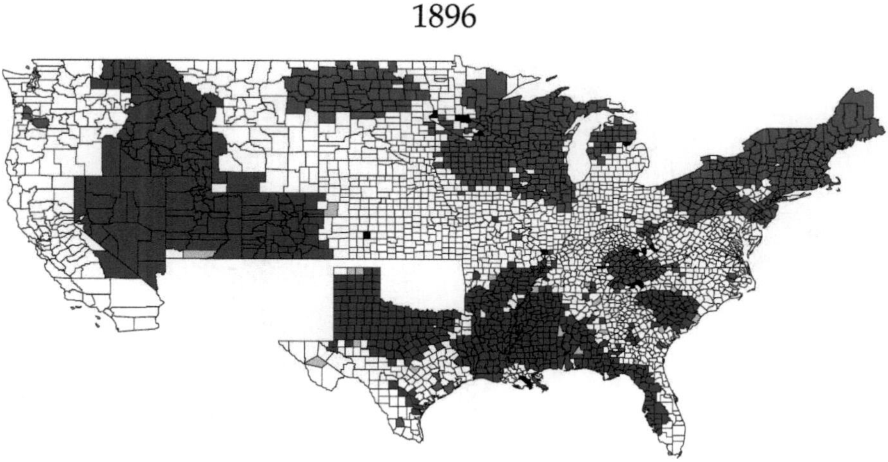

Fig. 2.20 The political geography of county-level voting in the 1896 presidential election

that ran from 1896 through 1928. And Wilson's two Democratic victories in 1912 and 1916 were only afforded by a divide between the progressive and conservative factions in the Republican Party. The Fourth Party System also saw the development of a strongly sectional party system, what Schattschneider (1960) called "the system of 1896."

The first election in the Fourth American Party System, the 1896 contest between Republican William McKinley and Democrat William Jennings Bryan, set the foundation for many of the issues that would dominate this party system. Bryan's agrarian, populist candidacy was made possible by the Panic of 1893, which occurred on the watch of the fiscally conservative Democratic President, Grover Cleveland. Bryan ran a rural-based populist candidacy opposed to wealthy owners of big businesses and the corruption in the cities and in support of the use of silver for currency (as reflected in his historic Cross of Gold speech at the Democratic National Convention that year). Bryan's insurgent candidacy was countered by McKinley's candidacy, which rejected the nativism of the Republican Party from the preceding decades in favor of a broader, big tent approach that would promote and harness the industrialism then sweeping the nation's big cities.

The 1896 map differs in some key ways from most of the maps during the preceding Third Party System. McKinley's strength throughout much of the Northeast is evident in Fig. 2.20. The solid red Republican regime included all counties in Connecticut, Maine, New Hampshire, Rhode Island, and Vermont, plus 91% of the counties in New Jersey, 87% of the counties in New York, 86% of the counties in Massachusetts, and 81% of the counties in Pennsylvania. McKinley also fared particularly well in the upper Midwest and upper Plains states, with 94% of the counties in Wisconsin, 76% of the counties in North Dakota, and 63% of the counties in Iowa.

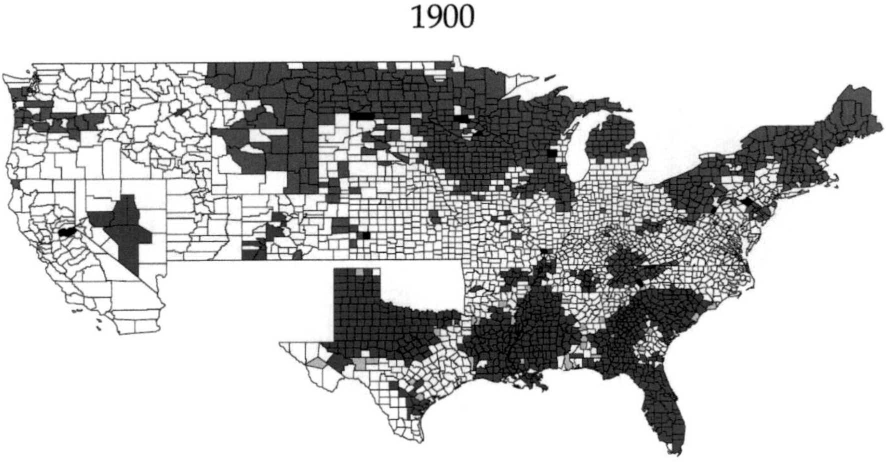

Fig. 2.21 The political geography of county-level voting in the 1900 presidential election

Bryan fared well in Mountain states where silver mining was located, as well as in the Democrats' base of the South. The blue spatial regime in Fig. 2.20 encompassed 93% of counties in Idaho and Utah, 82% of counties in Nevada, and 78% of counties in Colorado. In the South, the blue Democratic regime included 100% of counties in Mississippi, 85% of counties in South Carolina, 75% of counties in Arkansas and Louisiana, 60% of counties in Florida, 55% of counties in Texas, and 54% of the counties in Alabama. In the end, however, the 1896 election proved to be a realigning election (Nardulli 1995; Darmofal and Nardulli 2010), replacing the close national elections of the Third Party System with Republican dominance in the Fourth Party System.

The 1900 presidential election featured a rematch of the 1896 contest, pitting now-incumbent Republican President William McKinley against Democrat William Jennings Bryan. Buoyed by a strong economy and the U.S. victory in the Spanish-American War, McKinley enjoyed a more comfortable Electoral College victory in 1900 than he did in 1896. Interestingly, despite the fact that the candidates remained the same, the political geography of the 1900 election differed in important respects from that of the 1896 election.

In contrast to 1896, McKinley enjoyed larger spatial regimes of support in the upper Plains, Rocky Mountains, and Pacific Northwest, as can be seen in Fig. 2.21. Bryan lacked the regime of support he enjoyed in the Mountain states in 1896 but enjoyed a more expansive regime in the South in 1900. In all, McKinley enjoyed red regimes of Republican support in 35 states while Bryan enjoyed blue regimes of Democratic support in 16 states. McKinley's average margin of victory in the red counties was 28.7 percentage points while Bryan's average margin in the blue counties was 58.5 percentage points. But while Bryan ran stronger than McKinley in his counties, McKinley won a broader range of states and with them a comfortable Electoral College victory.

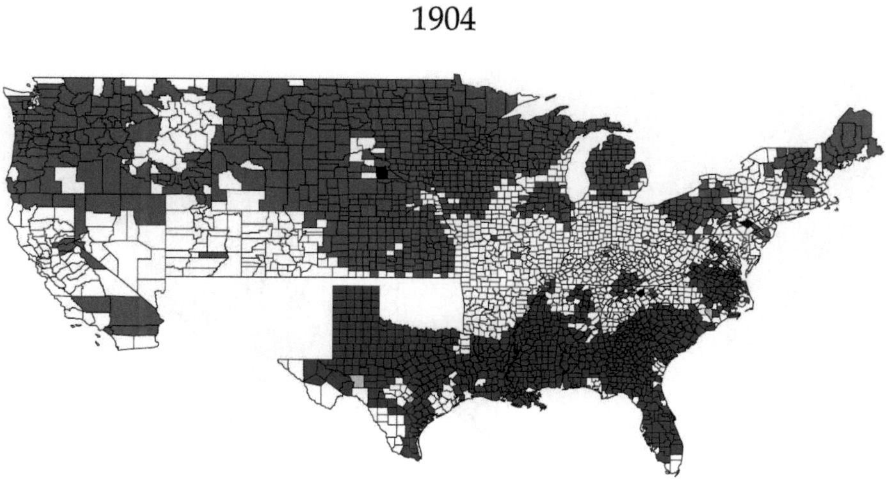

Fig. 2.22 The political geography of county-level voting in the 1904 presidential election

William McKinley was subsequently assassinated in 1901 and succeeded by his Vice President, Theodore Roosevelt. The 1904 election featured a contest between the now incumbent President Roosevelt and his Democratic opponent, Alton B. Parker. Parker was a marked departure from William Jennings Bryan, the Democratic nominee in the two preceding elections. A Gold Democrat, Parker instead marked a return to the more conservative policies of Bryan's Democratic predecessor, Grover Cleveland (Sanders 1999). Roosevelt would end up winning the 1904 election in one of the largest Electoral College and popular vote landslides up to that time.

Figure 2.22 presents the political geography of the 1904 election. As can be seen, the conservative Democratic nominee didn't enjoy the same support in Western, silver mining locations that Bryan had. But Parker's more conservative candidacy broke up some of the Northeastern Republican regime that had been so visible in the 1896 and 1900 elections. And Parker enjoyed an enlarged blue Democratic regime that now encompassed most of the South.

Rather than the Northeast and upper Plains and Mountain state Republican regimes of the two McKinley-Bryan elections, the 1904 election instead saw a breakup of the solid Northeastern Republican regime (recognizing that there were still areas of strong Republican support in the region) and a shift westward into the lower Plains and Pacific Northwest. In all, 29.7% of counties reflected the red Republican pattern while 25.8% of counties represented the blue Democratic pattern. Roosevelt's average margin of victory in the former counties was 45.8% while Parker's in the latter was 61.3%. In the 44.1% of counties that were spatially uncorrelated with their neighbors, Roosevelt enjoyed an average margin of victory of 6.4 points.

1908

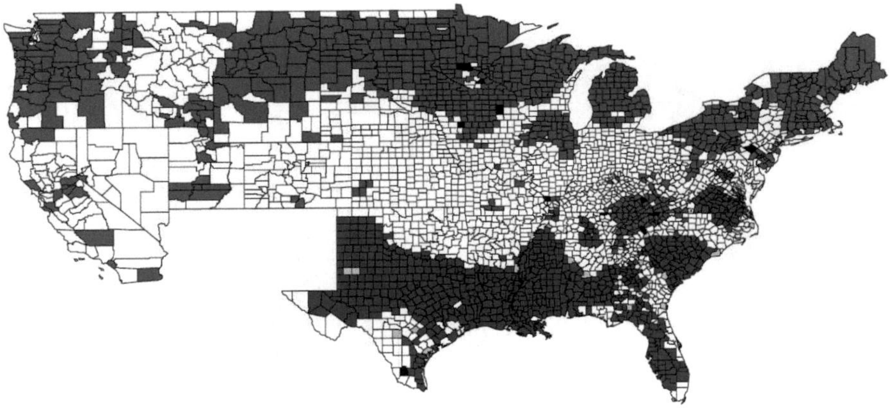

Fig. 2.23 The political geography of county-level voting in the 1908 presidential election

The 1908 election saw the third and final presidential candidacy for Democrat William Jennings Bryan. Bryan ran this time against Theodore Roosevelt's Secretary of War, William Howard Taft. Although Bryan fared relatively more poorly than he had in his two previous bids, he still was closer in both the Electoral College and the popular vote to the winner Taft than the 1904 nominee Parker had been to Roosevelt.

Figure 2.23 presents the 1908 election's political geography. As can be seen, with the Democrats' nomination of the populist Bryan as opposed to the more conservative Parker, the Republicans' spatial regime of support in most of the Northeast reappeared in 1908. Bryan's populist candidacy did, however, eliminate the Republican regime in the Southern plains from 1904. The Republicans' support on the West Coast from 1904 continued in 1908 as did Bryan's near complete domination of the South. As had been common in most elections, where the Democrat ran strong, he ran particularly strong, averaging much larger margins than Taft did in his red bastions of support. But Taft won where he needed to and the strongly sectional Fourth Party System continued to favor the Republicans, who were more popular than the Democrats in the populous Northern states.

The 1912 presidential election featured a three-way contest between incumbent President William Howard Taft, Democratic nominee and New Jersey Governor, Woodrow Wilson, and Progressive Party nominee and former President Theodore Roosevelt. Roosevelt was displeased with the conservative trajectory of his hand-picked successor, Taft, and after having been defeated for the Republican nomination in 1912, ran as a third-party candidate, dooming Taft's chances for re-election. Wilson ended up winning in an Electoral College landslide with 435 electoral votes to Roosevelt's 88 and Taft's 8. Wilson would end up being the only Democratic president between Grover Cleveland and Franklin Delano Roosevelt.

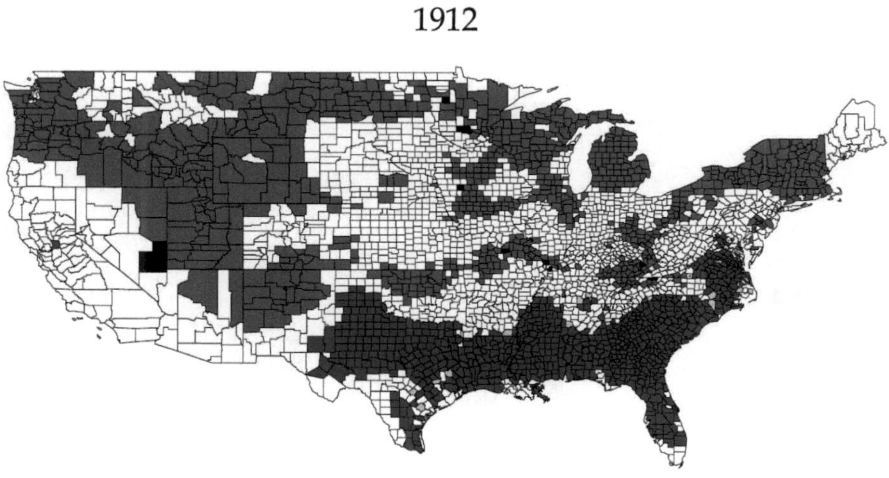

Fig. 2.24 The political geography of county-level voting in the 1912 presidential election

The political geography of the Democratic and Republican competition in 1912 is presented in Fig. 2.24. As we can see, despite the presence of a strong third-party candidate, the relative areas of Democratic and Republican strength continued as they had in preceding elections. Once again, the Democratic nominee enjoyed support from the reliably solid Democratic South. Wilson enjoyed an average margin over Taft in these counties of 74.2 points. The areas of relative Republican strength remained in the Northeast, the upper Midwest, and the West. Wilson's margin over Taft in these counties was a mere 1.4 points. And in the spatially uncorrelated counties, Wilson enjoyed an average margin over Taft of 26.8 points.

The 1916 presidential election featured a contest between incumbent Democratic President Woodrow Wilson and Republican nominee (and former Supreme Court Justice) Charles Evans Hughes. The election turned out to be one of the closest contests involving an incumbent president in American history, with Wilson winning the Electoral College 277–254 over Hughes.

Figure 2.25 presents the political geography of the 1916 election. As we can see, the election reflected the strongly sectional Fourth Party System. The blue Democratic spatial regime could be found in each of the 11 states of the former Confederacy. Equally tellingly, it existed in only two other states, Oklahoma and New Mexico, neither of which had been a state at the time of the Civil War. Conversely, the red Republican regime was present in only three of the former Confederate states, North Carolina, Tennessee, and Virginia. And it was present in 30 other states outside of the former Confederacy. In short, while the bloody shirt was no longer being waved with as much strength as in the Third Party System, the legacy of the Civil War continued to shape politics well into the Fourth American Party System.

The dominance of the Democratic Party in the blue regime can be seen from Wilson's average margin of victory in these counties. Wilson beat Hughes, on

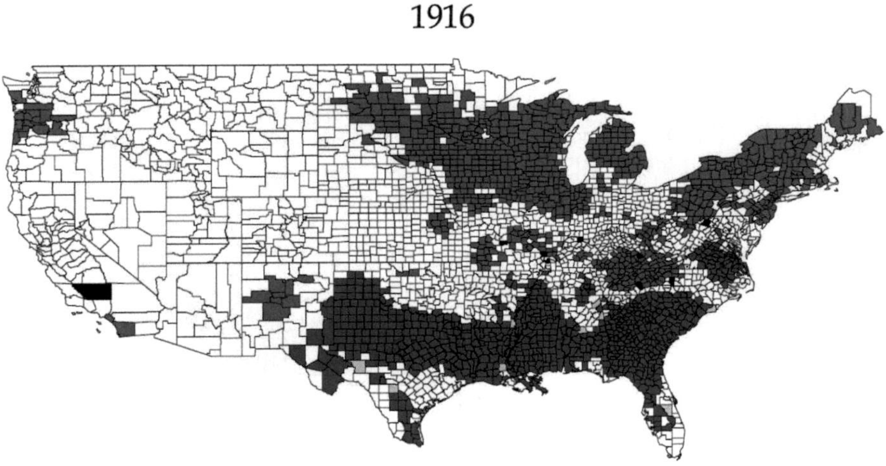

Fig. 2.25 The political geography of county-level voting in the 1916 presidential election

average, by 75 percentage points in these counties. In contrast, Hughes' average advantage over Wilson in the red counties in Fig. 2.25 was only 13.8 percentage points. In the uncorrelated counties, Wilson beat Hughes by an average of 16 percentage points. Wilson's victory in 1920 would be the last by a Democratic candidate in the Fourth Party System.

"He Kept us Out of War" was a Democratic campaign slogan in 1916. Early in 1917, however, the United States did enter World War I under Wilson, a decision that would have profound effects on the 1920 presidential election. The 1920 contest was a race between two Ohioans, Republican and Ohio Senator Warren G. Harding vs. Democrat and Ohio Governor James M. Cox. Eager to put the now-completed war behind them, American voters sought a "return to normalcy" in 1920, and favored Harding over Cox in one of the biggest landslides in American presidential election history (Dean 2004). Harding amassed 404 Electoral College votes to Cox's 127 and he defeated Cox in the popular vote by more than 26 percentage points.

And yet, as Fig. 2.26 shows, the political geography of the landslide 1920 election was similar in many regards to the geography of the close 1916 contest. Once again the Democratic candidate's blue regime of support extended across the states of the former Confederacy. And once again the Republican's support was focused outside of the old Confederacy. The principal difference between the two elections lies in Harding's regimes of support in the Upper Plains, Mountain states, and West Coast, which were largely absent in the 1916 election. Still, the predominant theme of the political geography of the 1920 election is that relative areas of Democratic and Republican strength do not differ as markedly between close and landslide elections as standard maps of political support, such as on election night newscasts, might suggest.

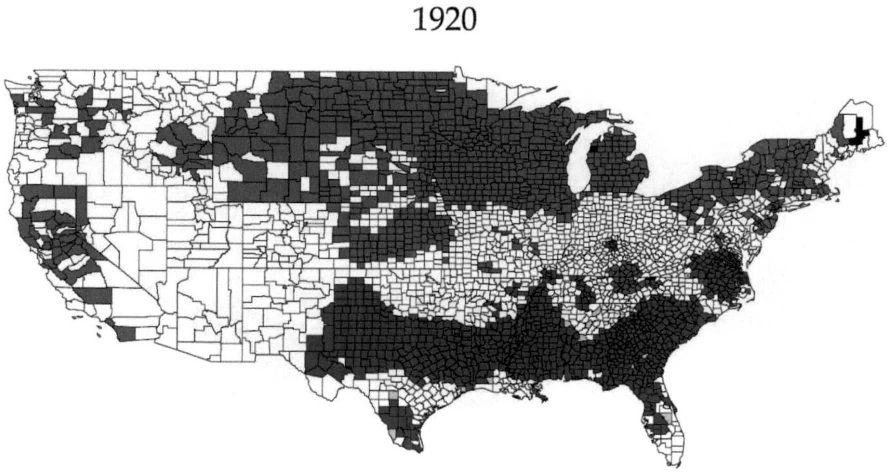

Fig. 2.26 The political geography of county-level voting in the 1920 presidential election

The 1924 election featured one of the more unusual contests in American history. The Republican candidate was incumbent President Calvin Coolidge, who succeeded Harding upon the latter's death in 1923. The Democratic Convention was perhaps the most unusual in American history, producing the most protracted convention balloting in history. Eventually, conservative West Virginia Democrat and former House member and Ambassador to the United Kingdom, John W. Davis, was nominated by the Democrats. Both faced a third-party challenge from Progressive Party candidate and Wisconsin Senator Robert M. La Follette.

Coolidge won the election in an Electoral College landslide, defeating Davis by an electoral vote count of 382 to 136. Davis carried the Solid Democratic South. Coolidge carried the rest of the country with the exception of Wisconsin, La Follette's home state, which La Follette carried. Coolidge also won a popular vote landslide, besting Davis by more than 25 percentage points.

The 1924 map in Fig. 2.27 once again highlights the Solid South of the Fourth Party System. Indeed, the counties in the blue Southern spatial regime are nearly identical to those in the regime from 1920. In all, Davis's average margin in the blue counties was 65 percentage points. In contrast to Davis's Solid South, Coolidge enjoyed strong support in the North. Just as Davis's area of support was similar to Cox's in 1920, so also was Coolidge's similar to Harding's from 1920. The principal differences in the Republican regimes in the two elections were a more consistent regime of Republican support in the Northeast and a diminished one in Wisconsin, LaFollette's home state. On average, Coolidge's margin of victory was 40.5 percentage points in the red counties, and 4.2 points in the uncorrelated counties. In all, Coolidge won a landslide victory over Davis in both the Electoral College and in the popular vote, where he bested Davis by 25.2 percentage points.

The final election of the Fourth Party System was the election of 1928. This election was a highly unusual one that produced changes in the standard political

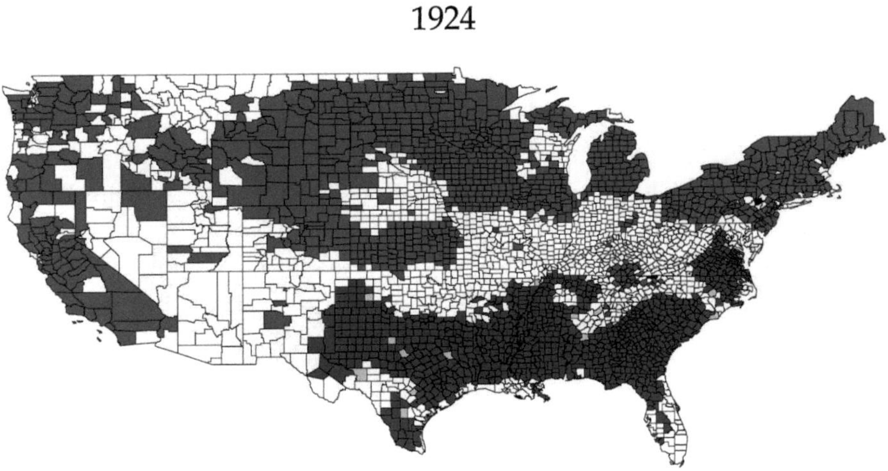

Fig. 2.27 The political geography of county-level voting in the 1924 presidential election

geography of voting during this party system due to the candidacy of Democrat Al Smith, the first Catholic nominee of a major party in U.S. history. With Smith on the ballot, anti-Catholicism was a major feature of the election. Smith's Republican opponent, former Commerce Secretary in the Harding and Coolidge administrations, Herbert Hoover, did better as a consequence of this anti-Catholicism in the South than Republican candidates had done in decades. Smith won only Alabama, Arkansas, Georgia, Louisiana, Mississippi, and South Carolina among the states of the Confederacy. At the same time, as a Catholic Governor of New York, Smith had more appeal to voters in the Northeast than the preceding Democratic candidates, Cox and Davis, and as a consequence, won both Massachusetts and Rhode Island. In the end, however, Hoover won the third Republican landslide in a row, more than quadrupling Smith's count in the Electoral College and winning the popular vote by over 17 percentage points.

The diminution of the Democrats' Southern regime of support is readily visible in Fig. 2.28. At the same time, however, so is the evaporation of the Republican regime in the North that had extended across the upper Midwest, Plains, and Rocky Mountain states in the preceding elections. Also diminished is the Republicans' regime in the Northeast. In short, the 1928 election was an unusually non-sectional election for the Fourth Party System.

The 1928 election was also noteworthy for the changes in average margins in this election in comparison to the preceding elections during this party system. Al Smith's average margin of 49.3 percentage points in the blue counties was drastically reduced from Davis's 65 point margin in 1924, Cox's 57.4 point margin in 1920, and Wilson's margins in the mid-70s in 1912 and 1916. At the same time, Hoover's average margin of 44.1 points was larger than Coolidge's in 1924 as well as most of the Republican margins in the Fourth Party System.

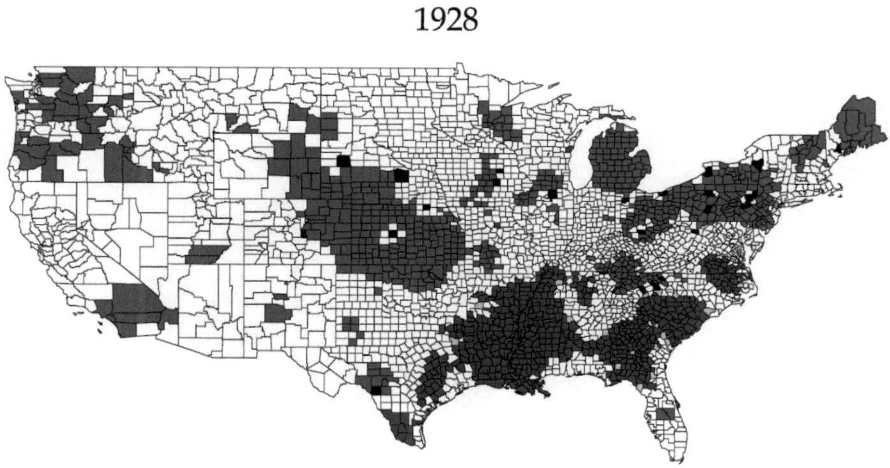

Fig. 2.28 The political geography of county-level voting in the 1928 presidential election

2.5.4 The Fifth American Party System

The stock market crash of October, 1929 and ensuing Great Depression brought an end to the Fourth American Party System. In its place arose the Fifth American Party System, featuring a strongly class-based politics for the next several decades. The Fifth Party System was forged through the leadership of Franklin Delano Roosevelt and his New Deal. While sectionalism continued in the Fifth System, the defining feature of the system was, again, its class-based politics with the Democratic Party representing working-class voters and Republicans generally being favored by upper-class voters. With the Democrats' emphasis on working-class concerns came an invigorated labor movement and the creation of large scale programs in the federal government to alleviate unemployment in the Great Depression. The Democratic Party also subsequently became associated with the country's successful prosecution of World War II, led by Roosevelt and his successor Harry S. Truman. Between them, Roosevelt and Truman would also lay the groundwork for the post-war international system while creating new expectations at home for federal involvement in the economy and in the creation of an economic safety net.

The first presidential election in the Fifth Party System was the 1932 presidential election, which took place three years after the stock market crash, in the depths of the Great Depression. The election pitted incumbent Republican President Herbert Hoover against Democratic nominee and New York Governor, Franklin Delano Roosevelt. FDR won the election in a landslide, carrying 42 states and 472 electoral votes in the Electoral College to Hoover's 6 states and 59 electoral votes. Roosevelt won the popular vote by nearly 18 percentage points.

Before turning to the political geography of the 1932 election, it is important to note that in a landslide election such as this, the blue and red regimes must be

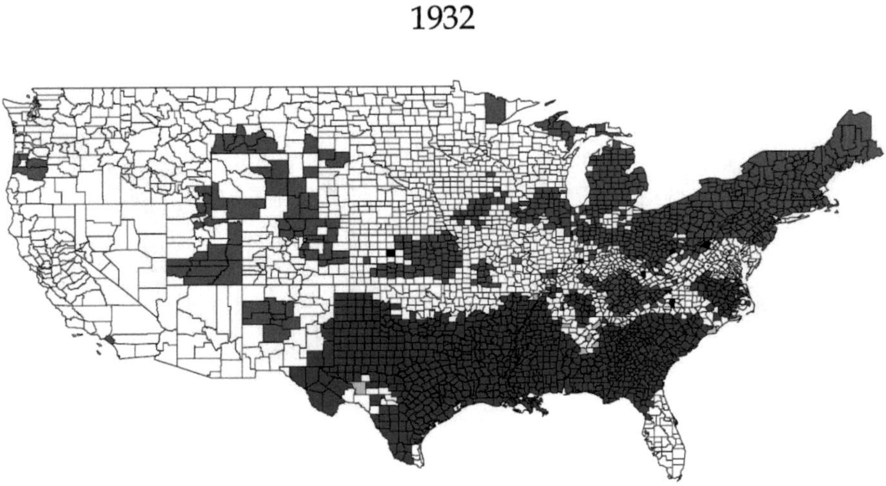

Fig. 2.29 The political geography of county-level voting in the 1932 presidential election

interpreted differently. In this case, rather than seeing the red regimes as strong Republican regimes, in a year like 1932 they were instead less Democratic regimes (based as the maps are on the Democratic proportion of the vote minus the Republican proportion of the vote). Indeed, Roosevelt's landslide was so strong that the mean margin of victory in the red spatial regimes in Fig. 2.29 was a .17 point Roosevelt victory margin. This slightly pro-Democratic margin was, of course, dwarfed by the margins for Roosevelt in the blue spatial regimes. The average margin in the blue counties in Fig. 2.29 was an 85 percentage point victory for FDR.

Looking at Fig. 2.29, we see that FDR's base of strongest support was the Solid South. A solid blue regime of strong Democratic support extended from west Texas to eastern South Carolina. In contrast, a spatially autocorrelated regime of less support for FDR extended throughout New England and westward into Ohio, Michigan, and Northern Indiana and Illinois. This was matched by the standard red regime in Eastern Tennessee as well as other red regimes in the western Plains and Rocky Mountain states. To get a better sense of the landscape of the 1932 election, we can also compare these areas to the spatially uncorrelated areas in white. Where Roosevelt's margin was only 0.17 percentage points in the red counties, it was 31.14 points in the spatially uncorrelated counties.

The landslide of 1932 thus served to shift the mean margin dramatically toward the Democrats from what it had been in the 1920s. And yet the relative areas of Democratic strength and less Democratic (more Republican) strength remained quite similar to what they had been in the Fourth Party System. For example, compare the political geography of the 1932 election to the geography of the 1924 election. While the outcome of the election was fundamentally different, the relative areas of strength remained largely the same. This again, is a recurring theme of the political geography of partisan voting in the United States—shifts in the mean that don't shift relative areas of strength. We will see this continue in later elections.

1936

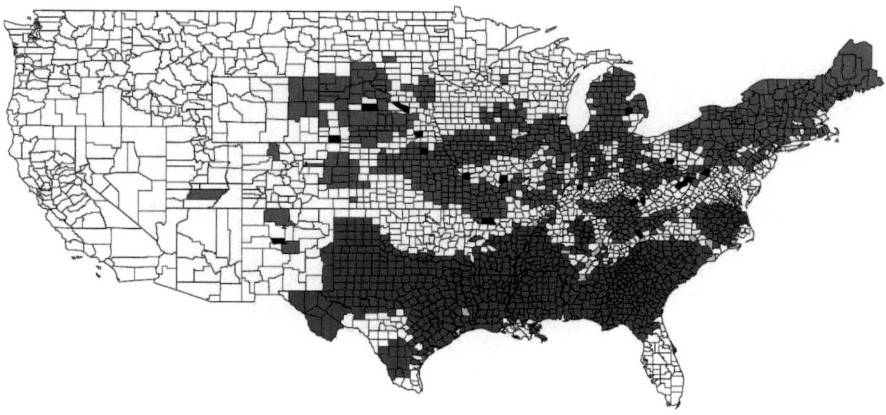

Fig. 2.30 The political geography of county-level voting in the 1936 presidential election

The 1936 election featured a contest between incumbent Democratic President Franklin Roosevelt and Republican nominee and Kansas Governor Alf Landon. In this election, *The Literary Digest* predicted a Landon victory based on a non-representative sample skewed toward the more affluent (Erikson and Tedin 1981). But on election day, FDR won the largest Electoral College landslide in American history, beating Landon in the Electoral College 523-8 and in the popular vote by over 24 percentage points. Landon would only win two states, Maine and Vermont, leading to the famous statement by Democratic National Committee Chair James Farley, "As Maine goes, so goes Vermont" (Lee 1982, 74).

The 1936 election map in Fig. 2.30 is remarkably similar to the 1932 map in Fig. 2.29. Again we see the solid Democratic South. And a red regime of decidedly less Democratic counties stretching from Maine to Iowa. In this case, these red counties were slightly Republican on average, with an average margin for Landon of 0.59 percentage points. Roosevelt's average margin in the blue counties of 82 percentage points was only slightly less than his average margin of 85 points in the blue counties in 1932. Once again, the spatially uncorrelated counties went overwhelmingly for FDR, this time by nearly 29 percentage points.

Franklin Roosevelt broke with the longstanding two-term precedent in running as the Democratic nominee for a third term in 1940. His opponent in the election was dark horse Republican nominee and businessman Wendell Willkie. Roosevelt easily won a third term, amassing over 400 electoral votes for the third time and winning the popular vote by nearly 10 percentage points.

The political geography of the 1940 election is presented in Fig. 2.31. Once again the solid Democratic South is present, stretching from west Texas to Virginia. Unlike the preceding two elections, the red spatial regime is not as present in the Northeast. A larger red regime is, however, visible in the Plains states. Most importantly, unlike

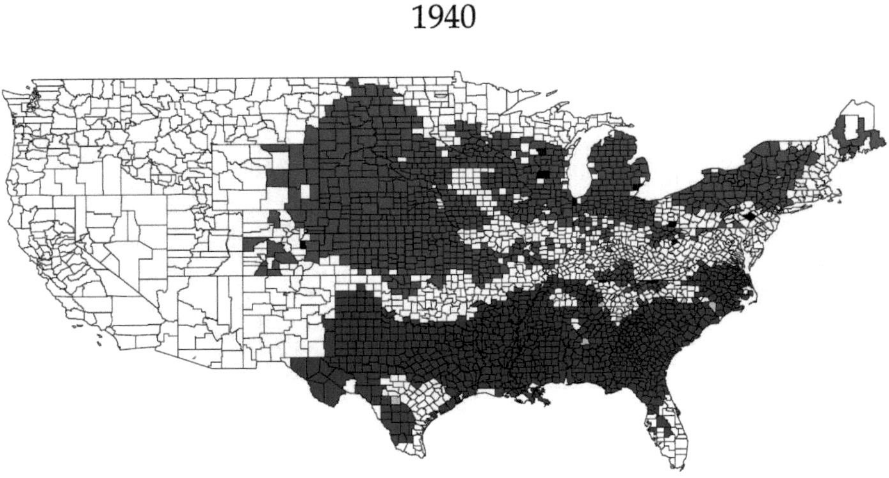

Fig. 2.31 The political geography of county-level voting in the 1940 presidential election

the two preceding Democratic landslides, the red counties in Fig. 2.31 do denote strong Republican voting. The average Wilkie margin in the red counties in this figure was nearly 18 percentage points. Roosevelt's margin in the blue counties was 74.9 percentage points, down slightly from the two preceding elections. The uncorrelated counties in Fig. 2.31 had an average Democratic margin of 14 points.

The 1944 presidential election featured incumbent Democratic President Franklin Roosevelt running for an unprecedented fourth term in office against Republican nominee and New York Governor Thomas Dewey. Although Dewey fared marginally better against Roosevelt than the three preceding Republican nominees, FDR still won an Electoral College landslide with over 400 electoral votes and won the popular vote by more than seven percentage points.

The map of the 1944 election in Fig. 2.32 is almost a photocopy of the 1940 election. In fact, 91.5% of counties in the United States were categorized in the same category on both maps. Once again, the South was solidly Democratic for Roosevelt. The red Republican areas in the Plains and Midwest that had supported Willkie in 1940 supported Dewey in 1944. Dewey did slightly better on average in the red counties in 1944 than Willkie had in 1940, with an average margin of victory of 23.6 points to Willkie's 17.9. And Roosevelt did slightly worse on average in the blue counties in 1944 than he had in 1940, with an average victory margin of 67.6 points in 1944 vs. 74.9 points in 1940. Overall, however, the story of the 1944 election was one of continuity with 1940. The similarities of the political geography in the two elections likely speaks to the importance of party loyalties that transcend specific elections.

Franklin Roosevelt died in April 1945 and was succeeded by his Vice President, Harry S. Truman. The 1948 election would feature a matchup between Truman as the incumbent president and Republican Thomas E. Dewey, the 1944 Republican nominee. Truman had to fend off challenges both from his left (the Progressive Party

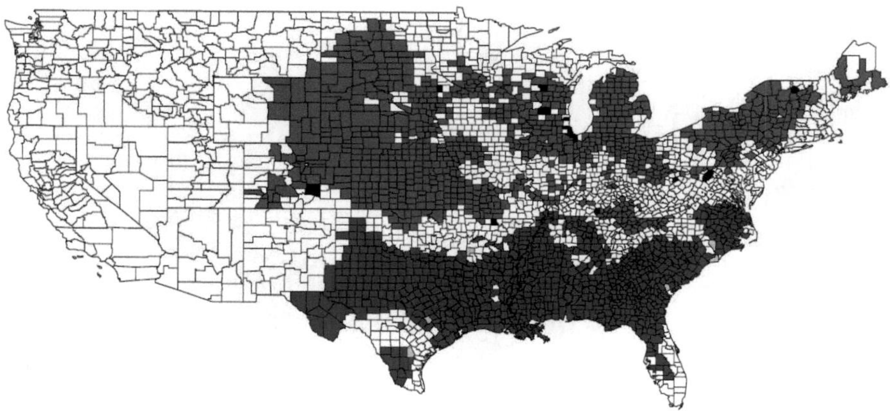

Fig. 2.32 The political geography of county-level voting in the 1944 presidential election

candidacy of Henry Wallace) and from his right (the Dixiecrat candidacy of Strom Thurmond). Able to rally Democrats back to his cause, he pulled off one of the greatest surprises in American presidential election history, winning over 300 Electoral College votes and defeating Dewey in the popular vote by four and a half points.

As we can see from Fig. 2.33, the South was not solidly Democratic for Truman. This reflects the strength in the region for Thurmond, who won four Southern states (Alabama, Louisiana, Mississippi, and South Carolina) plus an electoral vote in Tennessee. Still, Truman's support outside of the Deep South states plus Georgia is evident in Fig. 2.33. Dewey enjoyed a broader regime of support in New England than he had in 1944 and replicated his success from that year in the Midwest and the Plains. Truman's average margin in the blue counties in Fig. 2.33 was 54.4 points, down considerably from FDR's margins. Dewey's average margin in the red counties was 21 points, down marginally from his 23.6 point average margin in 1944.

While able to run for reelection in 1952, Truman was also deeply unpopular by this time and dropped out of the nomination contest early. The eventual Democratic nominee would be Illinois Governor Adlai Stevenson. His Republican opponent was former World War II General and national hero Dwight David Eisenhower. Eisenhower would win an easy landslide in 1952, ending a 20 year run of Democratic presidents.

The map of the 1952 election in Fig. 2.34 differs little in its general contours from those in the preceding 20 years. Once again, there is the solid Democratic South. Of the nine states that Stevenson carried in the election, seven were in the South (with the remaining two being Kentucky and West Virginia). And again we see the Republican strength in the Northeast, upper Midwest, and Plains states. Eisenhower was, however, able to extend the Republican regime to the Mountain states in 1952.

An interesting feature of the 1952 election is that for the first time since 1840, when William Henry Harrison's average margin was larger than Martin Van Buren's,

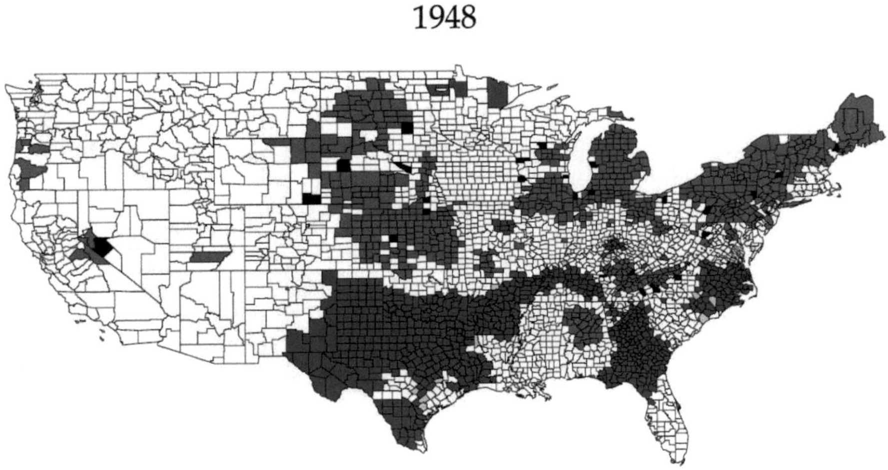

Fig. 2.33 The political geography of county-level voting in the 1948 presidential election

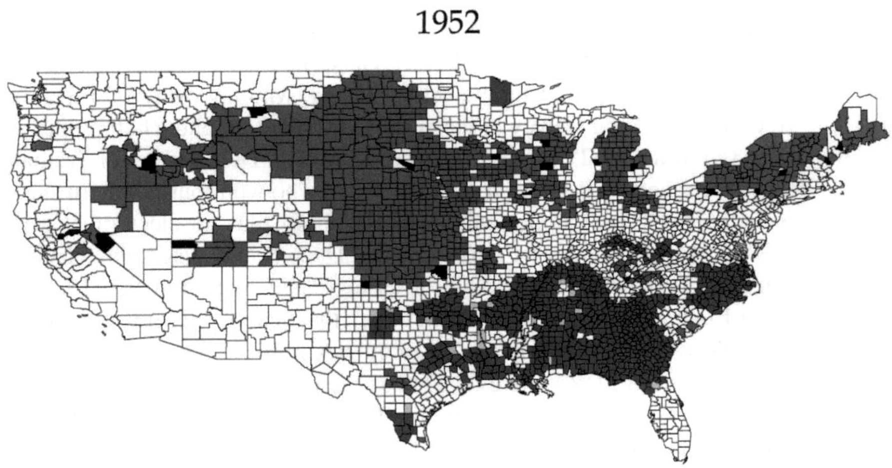

Fig. 2.34 The political geography of county-level voting in the 1952 presidential election

the average Democratic margin in the blue counties was smaller than the average margin for their opponent in the red counties. Stevenson won by an average of 35.8 points in the blue regimes, the lowest average Democratic margin to that point. Eisenhower's average margin was 42.7 points. And his average margin in the spatially uncorrelated counties was 14.3 points. Ike's victory in 1952 was both broad and deep.

The 1956 presidential election featured an unusual rematch from the election four years earlier. And the outcome of the 1956 election differed little from 1952's result. Once again Eisenhower defeated Stevenson in an Electoral College landslide, winning more than 400 electoral votes. Eisenhower also expanded his popular vote victory from just under 11 points in 1952 to over 15 points in 1956.

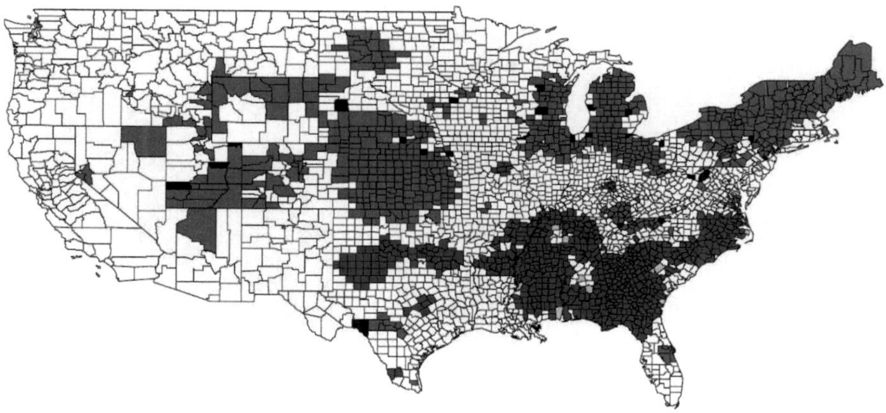

Fig. 2.35 The political geography of county-level voting in the 1956 presidential election

Despite the identical candidates and similar Electoral College results, the 1956 map in Fig. 2.35 bears some small differences from the 1952 map. The blue Democratic regime is much less present in Louisiana in 1956, which Eisenhower carried that year. Eisenhower's red Republican regime was stronger in New England in 1956 than it had been in 1952 while it was weaker in the upper Plains.

On average, Stevenson's support was stronger in the blue counties in 1956 than it had been in 1952. He had an average margin of 38.4 points in these counties in 1956 vs. 35.8 points in 1952. And Eisenhower's support in his red counties was slightly weaker in 1956 than it had been in 1952, at 39.7 points vs. 42.7 points in the election 4 years earlier.

The 1960 presidential election featured Massachusetts Senator and Democratic nominee, John F. Kennedy, against Eisenhower's Vice President, the Republican nominee, Richard M. Nixon. Kennedy and Nixon faced off in one of the closest presidential elections in terms of the popular vote, with Kennedy winning by just over 100,000 votes. The election was not as close in the Electoral College, however, with Kennedy gaining 303 electoral votes to Nixon's 219.

The map of the 1960 election is presented in Fig. 2.36. Here we see a pattern that would develop in the subsequent Sixth Party System: a much less sectional political geography. The solid South was not so solidly in the camp of Kennedy, the first successful Catholic nominee. But the north was not as consistently red as it had been in previous elections either. In fact, pockets of Democratic strength could be found well above the Mason-Dixon line, while pockets of Republican strength could be found in unexpected places in the South like Florida, South Carolina, and Texas.

Kennedy's average margin in his blue counties of 35.4 points outpaced Nixon's average margin in his red counties of 31.1 points. But in the uncorrelated counties, Nixon enjoyed an average 5.2 point margin. Beyond its closeness, however, the most important geographic feature of the 1960 election was its non-sectional nature

1960

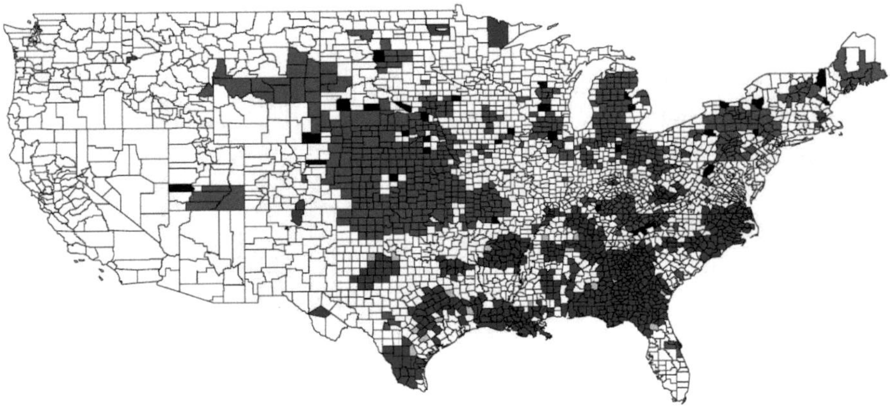

Fig. 2.36 The political geography of county-level voting in the 1960 presidential election

in which the South and North were no longer so dichromatically pitted against each other. Again, this would foreshadow the political geography of the Sixth Party System.

Following Kennedy's assassination on November 22, 1963, his Vice President Lyndon Baines Johnson became President. Johnson would run as the incumbent President in 1964 against Republican nominee and conservative Arizona Senator Barry Goldwater. Johnson defeated Goldwater in one of the biggest landslides in American presidential election history, winning in the Electoral College by a count of 486-52 and winning the popular vote by more than 22 percentage points.

The political geography of voting in the 1964 election is presented in Fig. 2.37. The most striking feature of the election map is something never previously seen in American election history: a regime of strong Republican support in the formerly Solid Democratic South. Goldwater enjoyed a regime of strong support from Louisiana (59% of its counties) eastward into Mississippi and Alabama (100% of the counties in both states) and on to portions of Georgia (55% of counties) and South Carolina (59% of its counties) as well as the panhandle of Florida (28% of Florida counties). The regime is likely a reflection of Johnson's support for the 1964 Civil Rights Act and Goldwater's opposition to it.

Meanwhile, extended blue spatial regimes could be found in the North where they had not existed before. Because the Solid Democratic South had crumbled, these Northern counties were now spatially autocorrelated at Democratic margins above the national average. In all, Johnson's blue Democratic regime could be found in 34 states: Arkansas, California, Colorado, Connecticut, Georgia, Iowa, Idaho, Illinois, Indiana, Kentucky, Massachusetts, Maine, Michigan, Minnesota, Missouri, North Carolina, North Dakota, New Hampshire, New Jersey, New Mexico, New York, Ohio, Oklahoma, Oregon, Pennsylvania, Rhode Island, South Carolina, Tennessee, Texas, Virginia, Vermont, Washington, Wisconsin, and West Virginia.

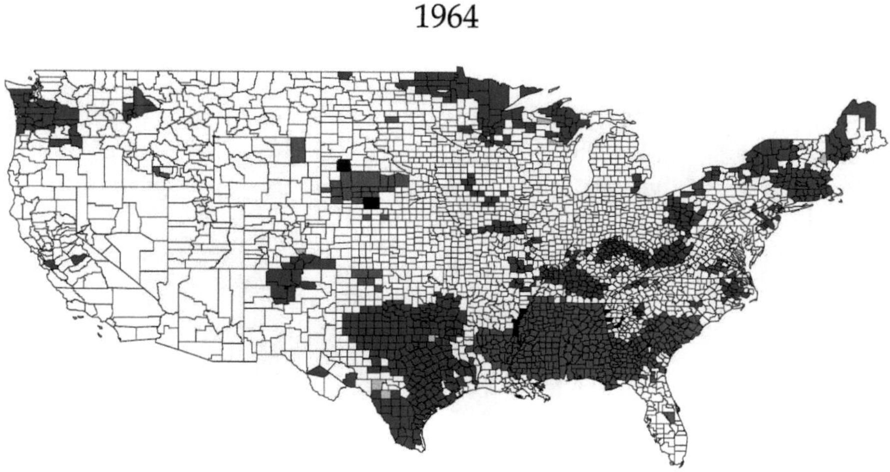

Fig. 2.37 The political geography of county-level voting in the 1964 presidential election

Where Johnson ran well, he ran very well, with an average margin of victory in the blue counties of 39.5 points. Goldwater was even stronger in the red counties, with an average margin of 46 points. But Johnson had a strong advantage in the spatially uncorrelated counties of 13.1 points on average.

2.5.5 The Sixth American Party System

There is considerable debate among political scientists as to whether we remain in the Fifth Party System or have moved into the Sixth Party System. We argue the latter, and specifically that the country moved into this new system in 1968, for several reasons. First, there have clearly been regional realignments in our view, with the formerly solid Democratic South becoming a Republican-dominated region and the Northeast becoming a Democratic-dominated region. Second, politics in recent decades has not been structured around the same social class cleavage that it was during the Fifth American Party System. Instead, race and cultural issues have played a different role than they did in the Fifth System. Third, while the Democratic Party was the dominant party at the presidential level in the Fifth System, winning all but two presidential elections from 1932–1964, the Democrats have not enjoyed this same dominance in the Electoral College since. In short, we believe the evidence is clear that the country moved into a Sixth Party System in roughly 1968, when the New Deal coalition broke apart over Vietnam and race.

The 1968 election featured a three-way race between Democratic Vice President Hubert Humphrey, Republican candidate and former Vice President Richard Nixon, and American Independent Party candidate and former Alabama Governor George

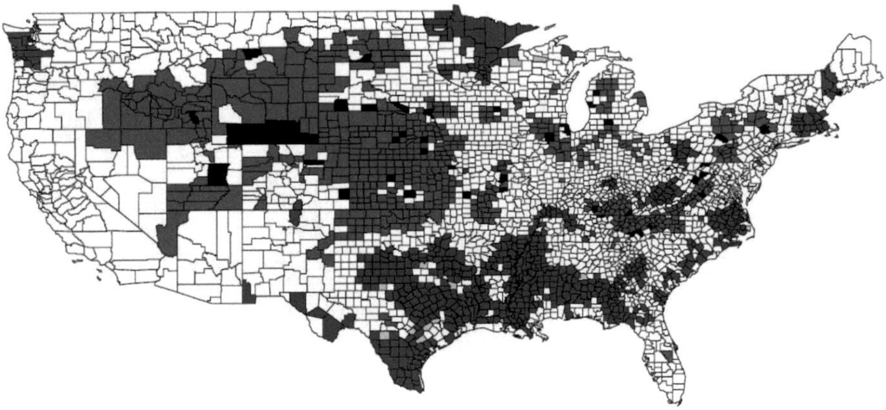

Fig. 2.38 The political geography of county-level voting in the 1968 presidential election

Wallace. Nixon won the election easily in the Electoral College, with 301 electoral votes to Humphrey's 191 and Wallace's 46, bringing to an end the Democrats' several decades domination of presidential elections during the Fifth Party System. Nixon's victory would be the first in a sequence of five Republican victories in the next six presidential elections. With Nixon's victory, predicated on a Southern strategy of appealing on cultural issues to both Southern whites and working-class whites in the North, Nixon would produce a realignment of American politics that still echoes through elections such as the 2016 Trump-Clinton contest.

Figure 2.38 presents the political geography of the 1968 election. Again, the patterns focus on competition between the two major parties, the Democrats and the Republicans. Here we see that Humphrey did well in comparison to Nixon in his native state of Minnesota, in portions of western Pennsylvania, West Virginia, New England, Washington state and much of the South. In contrast, Nixon performed well in portions of New York, central Pennsylvania, Ohio, Indiana, Illinois, Michigan, and a string of counties running from Virginia through North Carolina, Tennessee, and Kentucky. Nixon also performed well in the Plains and Rocky Mountain states. In all, Humphrey defeated Nixon by an average of 14.8 points in his blue strongholds in Fig. 2.37, but Nixon beat Humphrey by an average of 32.3 points in his red strongholds. Nixon beat Humphrey by an average of 8.9 points in the spatially uncorrelated white counties in Fig. 2.37.

The 1972 presidential election pitted incumbent Republican President Richard Nixon against Democratic nominee and South Dakota Senator George McGovern. McGovern ran on a liberal anti-war platform and lost the backing of many traditional, less liberal Democrats in his campaign against a popular incumbent president. Nixon ended up winning the 1972 election in one of the greatest landslides in American history, beating McGovern in the electoral vote by a margin of 520-17 and winning the popular vote by more than 23 percentage points.

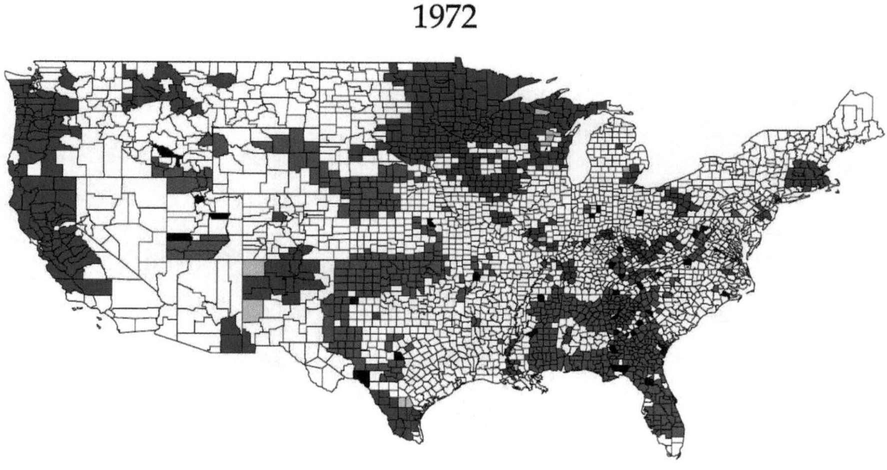

Fig. 2.39 The political geography of county-level voting in the 1972 presidential election

The political geography of the 1972 election is presented in Fig. 2.39. As can be seen from this figure, gone was the traditionally Solid Democratic South, replaced by a regime of strong Republican support in Alabama, Arkansas, Florida, Georgia, Louisiana, Mississippi, North Carolina, South Carolina, and Virginia. Another Republican regime could also be found in southern Kansas, extending down into Oklahoma and Texas. Stronger than the national average regimes of Republican support could also be found in Colorado, Idaho, Indiana, Kentucky, Missouri, New Mexico, Pennsylvania, Tennessee, Utah, and Wyoming. In all, Nixon averaged an astounding 57.9 point margin in the red counties, the largest average margin of victory for a Republican candidate in such counties since Abraham Lincoln's reelection in 1864.

In interpreting the blue regimes in Fig. 2.39, it is important to recall from the 1932 example that such regimes in a landslide election do not necessarily represent regimes of strong support for the losing party, but instead simply represent regimes of lower than average support for the party winning the landslide. Indeed, in the 1972 case, the average margin in the blue regimes was a Nixon victory of 11.5 points. Here we can see regimes of lower than average support for Nixon in the upper Midwest, Massachusetts (the only state that McGovern carried) and the West Coast (the latter perhaps foreshadowing cultural cleavages that have shaped recent presidential elections).

Richard Nixon would eventually be forced to resign in the wake of the Watergate scandal in August, 1974. His Vice President at the time, Gerald R. Ford, would run as the Republican nominee and incumbent President in 1976. His Democratic opponent was dark horse candidate and former one-term Governor of Georgia, Jimmy Carter. Carter would win the 1976 election in a close contest.

Figure 2.40 presents the map of the 1976 election. As the first successful presidential candidate from the Deep South, Carter was able to piece together some

1976

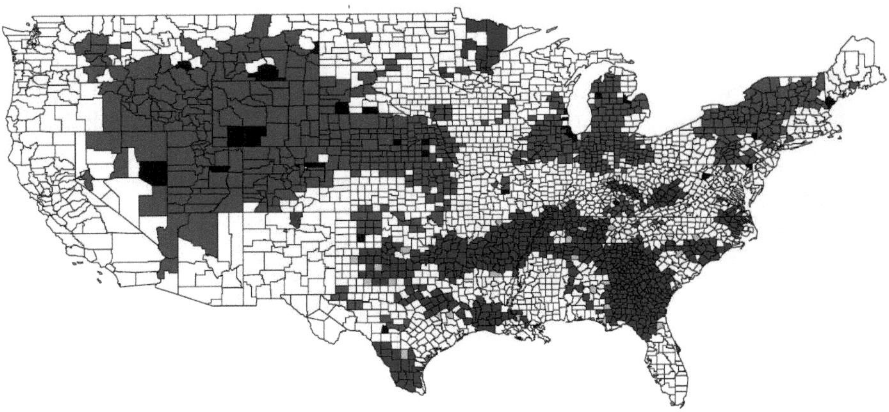

Fig. 2.40 The political geography of county-level voting in the 1976 presidential election

of the old blue Southern Democratic regime. Though perhaps in a harbinger of things to come, this blue spatial regime missed significant portions of Louisiana, Mississippi, Alabama, Florida, South Carolina, North Carolina, and Virginia. While Carter won all of these states except Virginia, the region was not a monolithic area of autocorrelated above average Democratic support.

Ford enjoyed regimes of red Republican support in New York, the Midwest, the Plains, and Mountain states. But Ford's margin in these red areas was not particularly strong, averaging only an 18.2 point margin. Conversely, Carter's margin in the blue counties was 37.4 points. And in the spatially uncorrelated counties, Carter enjoyed a small lead, on average, of 3.6 points.

The 1980 election featured a contest between the now-incumbent Democratic President Jimmy Carter and his Republican opponent, former California Governor Ronald Reagan. Mired in economic difficulties and facing a hostage crisis in Iran that had its one year anniversary on election day, Carter lost in an Electoral College landslide to Reagan. Reagan defeated Carter in the Electoral College, 489 to 49, winning 44 states and ushering in the Reagan Revolution.

And yet, despite two fundamentally different elections, one a close contest in which he was victorious and the other a landslide he lost, Carter's area of relative support in the South was little changed between the two elections. Likewise with Minnesota. This is clear from Fig. 2.41. This repeats a recurring pattern with the political geography of presidential voting: even when the mean support for the parties shifts, the *relative* areas of partisan support can stay quite similar.

On the Republican side of the ledger, Reagan repeated Ford's success in the Plains and Mountain states. He did not, however, enjoy the red spatial regime of relative support in the Northeast that Ford had in 1976. Overall, Reagan's average margin of victory in the red counties was 41.8 points. Carter's average margin in the blue counties was 17.6 points. And Reagan enjoyed a 13.4 point advantage in the spatially uncorrelated counties.

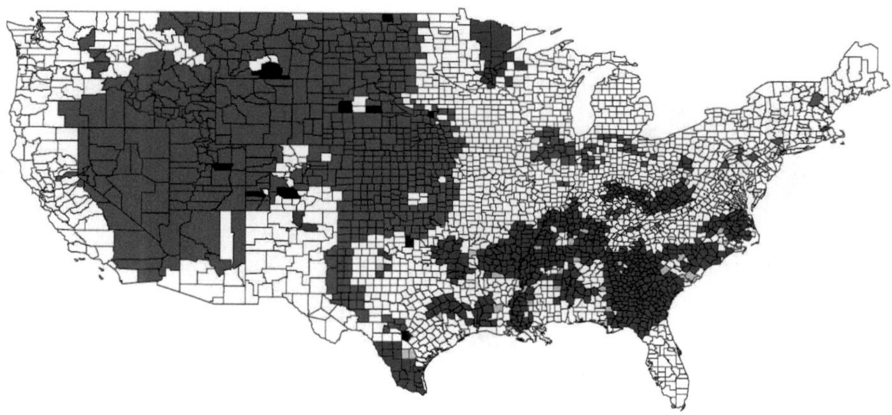

Fig. 2.41 The political geography of county-level voting in the 1980 presidential election

By 1984, Reagan was the incumbent Republican President, presiding over a recovering economy and running on an optimistic campaign theme of "morning in America." His Democratic opponent in 1984 would be Carter's Vice President, Walter Mondale. Reagan would win in a landslide, with 525 electoral votes to Mondale's 13 and winning the popular vote by over 18 percentage points. Mondale won only his home state and the District of Columbia.

The political geography of the 1984 election is presented in Fig. 2.42.[9] Again, the blue areas in this landslide do not represent areas of strong Mondale support (his average margin in these counties was only 0.05 percentage points) but instead of weaker Reagan support. And the red areas are areas of even stronger than the landslide average support for Reagan. (Reagan's average margin in the red counties was 50.2 points and his margin in the uncorrelated counties was 25.5 points).

Mondale, hailing from Minnesota, did not enjoy the relative area of support in the South that his Southern predecessor Carter had. Still, he was able to keep Reagan's margins down in the Black Belt areas of the South and along the Mississippi Delta. Mondale also was able to limit Reagan's support in other areas of the country, such as eastern Ohio and western Pennsylvania, Iowa, and his home state of Minnesota. Reagan, meanwhile, once again enjoyed the stronger than average support in the Plains and Mountain states. He also expanded his regime from 1980 in western and rural Ohio and added a pocket of strong support in central Florida.

The 1988 presidential election featured a matchup between Reagan's Vice President George H.W. Bush and the Democratic nominee, Massachusetts Governor Michael Dukakis. Ahead by 17 points following the Democratic National Convention, Dukakis saw this lead disappear as the campaign progressed. On election day,

[9]Figure 2.42 was published previously as Figure 4.2 (page 57) in Darmofal (2015). Reprinted with permission.

1984

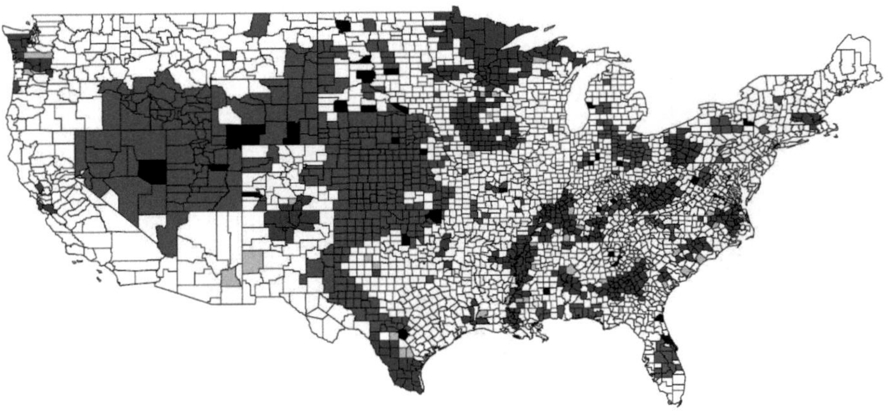

Fig. 2.42 The political geography of county-level voting in the 1984 presidential election

1988

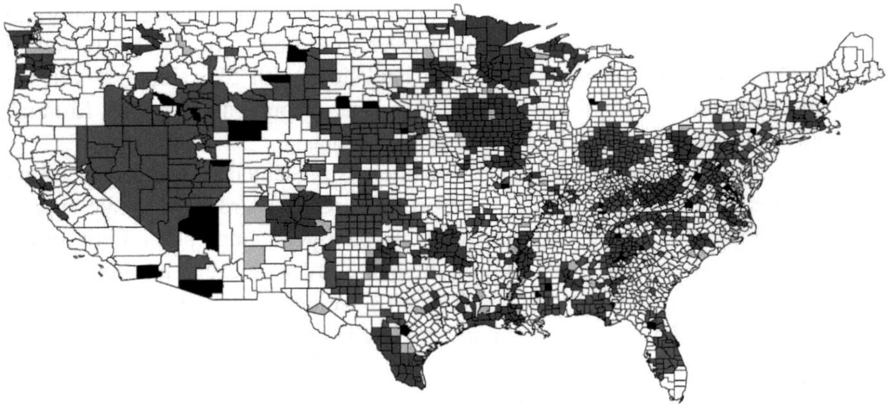

Fig. 2.43 The political geography of county-level voting in the 1988 presidential election

Bush won the third resounding Republican victory in a row, although Dukakis won a set of states in the Northern tier of the country.

Figure 2.43 presents the political geography of the 1988 election. The most telling feature of this geography is the lack of any strong regional strengths. Instead, the regimes, much more than preceding elections, represent pockets of relative Republican and Democratic strength. Dukakis fared well in his home state of Massachusetts, in eastern Ohio and western Pennsylvania, in West Virginia, in the farm country of Iowa, western Wisconsin and Illinois, the Upper Peninsula of Michigan and Minnesota, in the Mississippi Delta, southern Texas, northern New Mexico, northern coastal California, and in the Pacific Northwest.

1992

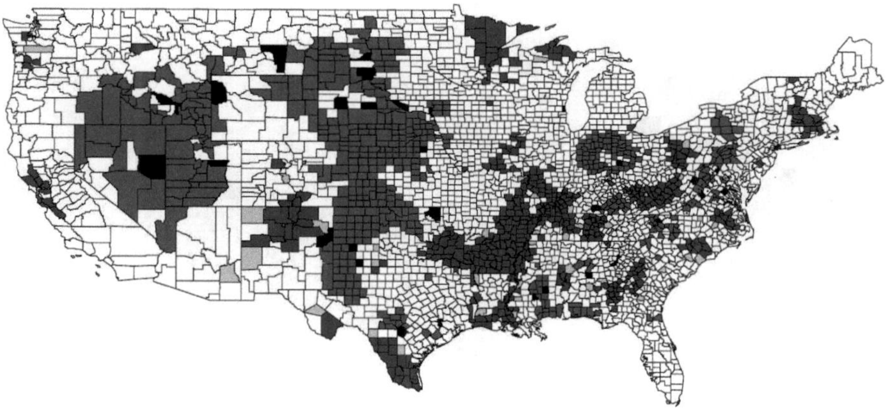

Fig. 2.44 The political geography of county-level voting in the 1992 presidential election

Bush, meanwhile, repeated Reagan's strength in central Florida and in western Ohio. He also fared particularly well in Indiana, portions of Appalachia, the Plains, and Nevada and Utah. Dukakis enjoyed an average margin of 11.7 points in the blue counties. But Bush enjoyed a much larger margin of 34.9 points in the red counties. He also led in the uncorrelated counties by an average of 12.6 points.

The 1988 election would prove to be the end point of Republican dominance in presidential elections. As stated earlier, Republican candidates won 5 of 6 presidential elections from 1968 through 1988. But Democrats would win the popular vote in 6 of the next 7 presidential elections from 1992 through 2016, while winning 4 of 7 in the Electoral College.

The 1992 election was thus a turning point in partisan strength in presidential elections. In this election, Arkansas Governor Bill Clinton faced off against incumbent Republican President George H.W. Bush and independent candidate and Texas billionaire Ross Perot during a time of weak economic performance. Clinton would win an Electoral College landslide of 370 electoral votes to Bush's 168.

Figure 2.44 presents the political geography of the 1992 election. Clinton's strength in his home state of Arkansas and up into Missouri and southern Illinois is evident as is his strength in Kentucky and West Virginia, eastern Ohio and western Pennsylvania, and portions of Georgia. Clinton also fared well in the traditional Democratic stronghold of Minnesota as well as in Massachusetts and Vermont.

Bush did well once again in western Ohio and Indiana as well as in the Plains and Mountain states. Clinton's margin in the blue counties was 22.2 points while Bush's margin in the red counties was 21.9 points. Clinton enjoyed a modest advantage of 0.4 points in the spatially uncorrelated counties.

By 1996, the economy was beginning to boom again. The incumbent Clinton faced off against the Republican nominee and former Kansas Senator Robert Dole

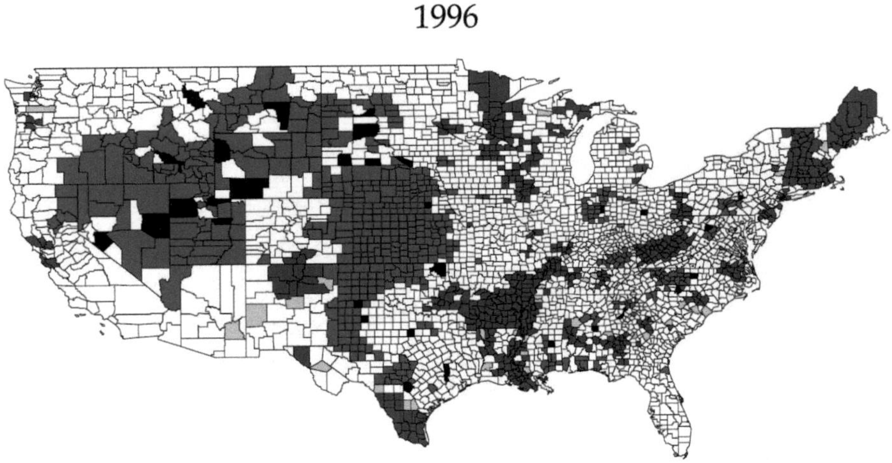

Fig. 2.45 The political geography of county-level voting in the 1996 presidential election

and, once again, Ross Perot, this time the candidate of the Reform Party. Clinton once again won an Electoral College landslide, beating Dole by a margin of 379 to 159.

Figure 2.45 presents the political geography of the 1996 election. There are definite similarities to the geography of the 1992 election. Once again Clinton fared particularly well in his home state of Arkansas, in Kentucky and West Virginia, in southern Texas, pockets of the South, and northern California. To this Clinton added a strong blue regime in New England.

Dole repeated the Republican regime in the Plains and Mountain states. But he lacked the areas of support in rural Ohio that had become common for Republicans in the preceding elections. While Clinton fared well in the blue counties with an average margin of 23.3 points, Dole enjoyed an even larger average margin of 27.9 points in the red counties. Clinton had a very modest edge of 0.2 points in the spatially uncorrelated counties.

The 2000 election featured a contest between Clinton's Vice President Al Gore and his Republican opponent, Texas Governor George W. Bush. Following Clinton's affair with Monica Lewinsky and his subsequent impeachment, Gore struggled particularly in separating himself effectively from Clinton's personal behavior while also being able to claim the economic and policy successes of the Clinton administration. Bush ran on a theme of change and restoring honor and integrity to the presidency. The election would end up being the closest in American history, with Bush winning a narrow Electoral College victory based on a 537 vote margin in Florida while Gore won the popular vote. Gore would eventually concede the election more than a month after election day, after the Supreme Court intervened in the *Bush v. Gore* decision to stop the Florida recount and effectively hand the election to Bush.

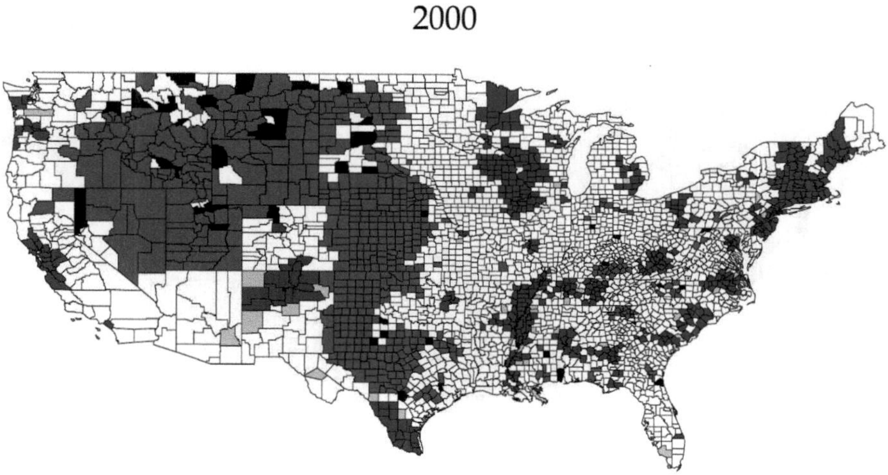

Fig. 2.46 The political geography of county-level voting in the 2000 presidential election

The political geography of the 2000 election is presented in Fig. 2.46.[10] As can be seen, the geography of this election was quite similar to the geography of the 1996 election. Again, Gore fared well along the Mississippi Delta, in the Upper Midwest, and in portions of New England and pockets of the South. Bush fared well in the Plains states, his home state of Texas, and the Mountain states.

Having won a narrow victory in 2000, Bush ran as the incumbent President in 2004 against Democratic nominee and Massachusetts Senator John Kerry. Occurring in the context of an increasingly unpopular war in Iraq, the election once again was close. In the end, Bush won a narrow Electoral College victory by winning the state of Ohio. Had Kerry carried the state, he would have won the election.

The political geography of the 2004 election is presented in Fig. 2.47. There are clear similarities between the geography of this election and preceding elections. Again, Kerry fared particularly well in the Mississippi Delta, in the Upper Midwest, and in New England. Bush repeated his strength in his home state of Texas, in the Plains states, and in the Mountain states. He also had more pockets of strength in the South than he had enjoyed in 2000. In all, however, the consistency of the pattern with those of preceding elections suggests a maturing party system settling into relatively stable areas of partisan strength.

The 2008 presidential election took place in the shadow of an unpopular war in Iraq and during a financial meltdown during the fall campaign. The election pitted Illinois Senator and Democratic nominee Barack Obama against Arizona Senator and Republican nominee John McCain. Obama would win the election in an Electoral College landslide of 365 to 173. In addition to reliably Democratic states, Obama won states that had not voted for a Democratic candidate in decades,

[10]Figure 2.46 was published previously as Figure 4.4 (page 59) in Darmofal (2015). Reprinted with permission.

2004

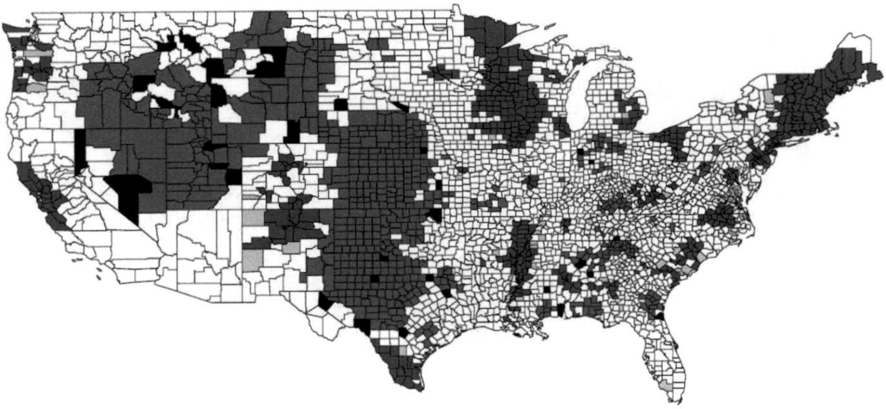

Fig. 2.47 The political geography of county-level voting in the 2004 presidential election

including North Carolina which last voted for a Democrat in 1976, and Virginia and (most unexpectedly) Indiana, which had not voted for a Democrat since 1964. Obama beat McCain in the popular vote 52.9% to 45.7%. With his election to the presidency, Barack Obama became the first African American president in the nation's history.

The political geography of the 2008 presidential election is presented in Fig. 2.48. As can be seen, although Barack Obama won the 2008 election in an Electoral College landslide with a majority of the popular vote, his relative areas of support were little changed from John Kerry's in 2004. Likewise for John McCain in comparison to Bush's 2004 support. Like Kerry, Obama performed well in New England, portions of the Southeast, the Mississippi Delta, the upper Midwest, and the West Coast. McCain, meanwhile, enjoyed the support in Texas, the Plains states, the Rocky Mountain states, and portions of the South that Bush had in 2004.

In all, the spatially autocorrelated blue counties in 2008 constituted 16.6% of counties, while the spatially autocorrelated red counties accounted for 20.24% of counties. Spatially uncorrelated counties accounted for 61.23% of counties. Obama's average margin of victory in his strongholds was 17.36 points, up markedly from Kerry's average margin of victory of 8.16 points in his strongholds in 2004. McCain's average margin of victory in the red counties was 47.5 points, down slightly from Bush's average margin of 52.42 points in his red counties in 2004. Even in the Democratic election year of 2008, the average margin of victory in the spatially uncorrelated counties was a pro-McCain margin of 13.86 points. This was, in part, a harbinger of the shrinking geographic base of the Democratic Party, something that would become more visible in the 2016 election.

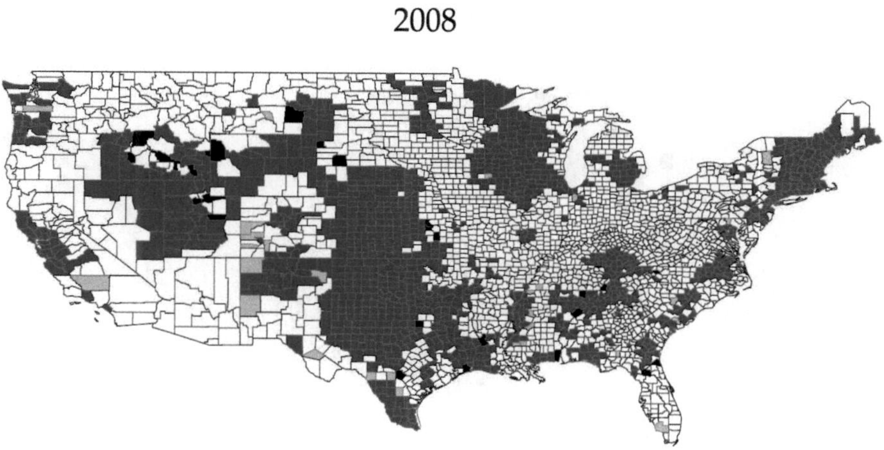

Fig. 2.48 The political geography of county-level voting in the 2008 presidential election

The 2012 presidential election featured a contest between incumbent Democratic President Barack Obama and Republican nominee and former Massachusetts Governor Mitt Romney. The election took place during a slowly improving economy following the Great Recession of 2008. Obama won reelection easily, defeating Romney in the Electoral College 332-206. Obama also became the first Democratic presidential candidate since FDR to twice win the popular vote by a majority, beating Romney 51.1% to 47.2% in the popular vote.

The political geography of the 2012 presidential election is presented in Fig. 2.49. Although both Obama's Electoral College and popular vote margins declined in 2012 in comparison to 2008, the map of the political geography of the 2012 election was little changed from 2008's map. Once again, Obama was strong in New England, the upper Midwest, portions of the Southeast, the Mississippi Delta, and on the west coast. Romney enjoyed the same area of relative support in Texas, the Plains states, and the Rocky Mountain states that McCain and previous Republicans had enjoyed.

As would be expected from the tighter margins in 2012, Obama's average margin of victory in the blue counties was down slightly in 2012 from 2008, at 14.88 points instead of 2008's 17.36. Romney's average margin in the red counties was up from McCain's, at 54.29 points in contrast to McCain's 47.50. The spatially uncorrelated counties in 2012 were markedly more Republican than in 2008, with an average Republican margin of 20.85 points up from 13.86 in 2008.

The 2016 presidential election proved to be one of the most unusual and unexpected in American history. The election featured a matchup between Barack Obama's former Secretary of State, former Senator from New York, and former First Lady, Hillary Clinton against businessman and star of NBC's The Apprentice, Donald Trump. Clinton sought to become the first Secretary of State to win the presidency since James Buchanan in 1856. In the most surprising election result

2012

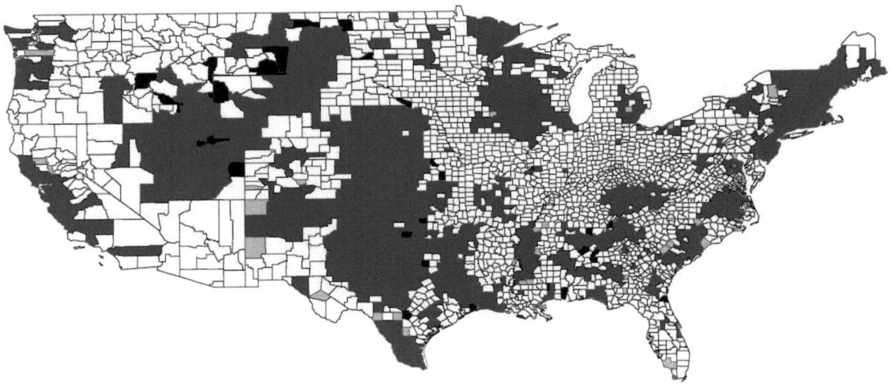

Fig. 2.49 The political geography of county-level voting in the 2012 presidential election

since Truman's defeat of Dewey in 1948, Donald Trump was elected president with an Electoral College win of 304 electors to Clinton's 227. Clinton won the popular vote by nearly three million votes, the largest such margin for a candidate who lost the electoral vote. The 2016 contest thus marked the second time in five elections that the popular vote winner lost in the Electoral College. Donald Trump became the first candidate elected president in the nation's history without previous experience in the government or military.

The political geography of the 2016 presidential election is presented in Fig. 2.50. The Clinton-Trump map does present some departures from the two Obama-era maps. While Clinton, like Obama, enjoyed strong areas of support in New England and portions of the Southeast, her support was reduced in the upper Midwest in comparison to Obama. In contrast, she enjoyed more support throughout California than Obama had as his coastal bastions of support spread to the inland regions of California. In contrast to the two preceding Republican presidential candidates, Trump enjoyed a broadened area of support in Appalachia. This was paired with the usual areas of Republican strength in Texas and the Plains with lesser support in the Rocky Mountain states than the two preceding Republican candidates had enjoyed.

Clinton's average margin in the blue counties in Fig. 2.50 was much reduced from Obama's average margins in his blue counties. Where he enjoyed average margins of 17.36 and 14.88 points, respectively, in 2008 and 2012 (already well below the average margins for his Republican opponents), Clinton's average margin in her blue counties was only 10.34 points. And while Clinton's average margin was smaller, Trump's average margin was larger than McCain's and Romney's. Where McCain enjoyed an average margin of 47.5 points in the red counties in 2008 and Romney enjoyed one of 54.29 in 2012, Trump's average margin of victory in the red counties in 2016 was a whopping 61.95 points. This was the largest average margin of victory for any Republican candidate in the red counties since Lincoln's defeat of

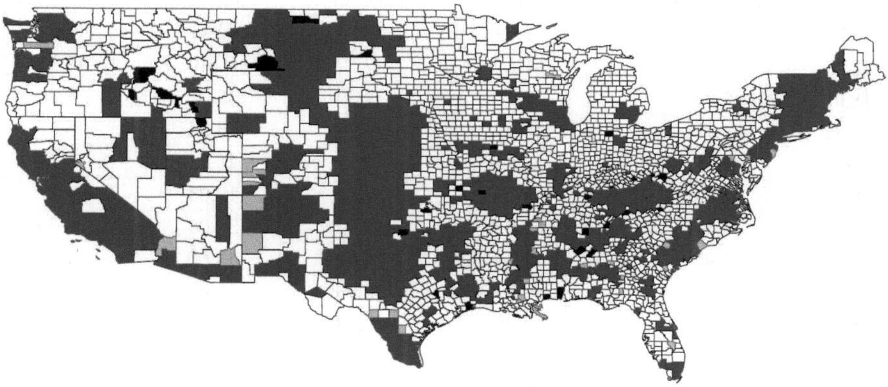

Fig. 2.50 The political geography of county-level voting in the 2016 presidential election

McClellan in 1864. And where there was an average Republican advantage of 13.86 points in the spatially uncorrelated counties in 2008 and of 20.85 in 2012, there was an average advantage for Trump of 32.43 points in the spatially uncorrelated counties in 2016. In short, Trump ran more strongly in his strongholds than previous Republican candidates and more strongly in the spatially uncorrelated counties than the preceding two Republican candidates while Clinton's margins in her strongholds were reduced from Obama's margins.

2.6 Conclusion

Since 2000, academics and popular observers alike have often viewed the country as being divided between Republican "red states" and Democratic "blue states." But is this (red state-blue state)' dichotomy the best way to understand the political geography of voting in the United States? And is the era since 2000 somehow a fundamentally different era from what came before it? The spatial analyses in this chapter argue that both of these common conceptions are, in fact, misperceptions that distort our understanding of the political geography of voting in the United States. This geography is, in fact, much richer, complex, and varied than popular commentaries would suggest (Tables 2.3 and 2.4).

We're able to shed new light in this chapter on the political geography of voting because we employed spatial diagnostics to identify, for the first time, the locations of like-voting Democratic and Republican regimes in the United States. We applied these spatial techniques to the entire period of mass voter participation in the United States since the advent of Jacksonian democracy in 1828. And in the process, we've

Table 2.3 Mean margins by Moran scatterplot category, 1828–2016

Year	Not significant	High-high	Low-low	High-low	Low-high
1828	23.52	75.99	−22.81	44.11	15.71
1832	41.51	91.57	−7.57	60.40	14.25
1836	5.94	56.44	−37.39	35.32	−8.26
1840	−4.71	40.99	−47.89	13.33	−15.63
1844	4.96	46.45	−30.15	22.53	−5.64
1848	3.46	51.29	−32.22	19.08	−3.86
1852	15.52	60.67	−15.92	32.04	4.39
1856	47.97	71.61	−18.62	48.60	31.75
1860	2.58	39.09	−38.99	16.00	−4.00
1864	−13.60	43.30	−64.47	1.04	−30.82
1868	−6.41	58.18	−52.88	18.06	−18.15
1872	−5.78	49.83	−49.61	11.83	−19.90
1876	9.95	64.55	−33.31	39.93	5.21
1880	8.06	61.82	−31.78	21.42	−4.72
1884	4.91	58.02	−25.70	19.25	−2.66
1888	5.38	62.22	−22.37	21.11	−7.03
1892	4.83	59.15	−36.59	24.60	−10.86
1896	8.92	67.19	−26.47	26.00	−4.59
1900	3.33	58.52	−28.71	15.18	−6.18
1904	−6.44	61.27	−45.81	2.00	−12.75
1908	1.96	61.80	−29.02	15.25	−5.05
1912	26.82	74.18	1.44	37.89	22.88
1916	15.97	75.01	−13.76	33.33	−3.00
1920	−11.30	57.40	−47.59	3.00	−14.43
1924	−4.17	65.00	−40.51	12.50	−11.00
1928	−15.11	49.29	−44.12	−1.36	−24.00
1932	31.14	85.01	0.17	44.86	23.70
1936	28.92	82.09	−0.59	38.24	18.79
1940	13.97	74.87	−17.86	26.50	13.00
1944	7.94	67.63	−23.60	20.23	−00.29
1948	6.08	54.44	−21.04	17.93	2.50
1952	−14.28	35.75	−42.71	−4.52	−20.57
1956	−12.99	38.40	−39.69	−4.46	−18.08
1960	−5.19	35.41	−31.14	6.03	−11.00
1964	13.08	39.48	−46.00	24.00	0.19
1968	−8.90	14.77	−32.25	−0.22	−15.00
1972	−34.89	−11.51	−57.85	−25.21	−41.33
1976	3.64	37.40	−18.20	1.29	−2.69
1980	−13.40	17.62	−41.82	1.55	−20.91
1984	−25.49	0.05	−50.21	−10.08	−33.86
1988	−12.55	11.67	−34.92	1.03	−21.61

(continued)

Table 2.3 (continued)

Year	Not significant	High-high	Low-low	High-low	Low-high
1992	0.38	22.15	−21.88	12.69	−9.09
1996	0.16	23.32	−27.89	11.94	9.22
2000	−16.00	10.39	−47.21	−0.70	−27.53
2004	−20.28	8.16	−52.42	−8.00	−30.68
2008	−13.86	17.36	−47.50	2.19	−29.50
2012	−20.85	14.88	−54.29	−3.39	−33.48
2016	−32.43	10.34	−61.95	−14.86	−44.26

Table 2.4 Percentage of counties in each Moran scatterplot category, 1828–2016

Year	% Not significant	% High-high	% Low-low	% High-low	% Low-high
1828	58.43	22.72	17.58	0.71	0.55
1832	45.60	27.57	25.50	0.74	0.59
1836	70.19	13.92	13.20	1.31	1.37
1840	71.68	13.56	12.72	1.08	0.96
1844	72.82	13.50	11.05	1.14	1.49
1848	68.39	15.47	14.23	1.24	0.67
1852	65.89	17.15	14.31	1.35	1.30
1856	50.13	26.21	23.25	0.22	0.18
1860	68.02	13.46	17.04	0.89	0.60
1864	67.91	13.06	16.27	1.97	0.80
1868	65.10	15.27	17.45	0.87	1.30
1872	63.08	16.68	18.40	0.69	1.15
1876	59.36	19.94	18.87	0.58	1.27
1880	57.43	19.99	21.04	0.45	1.09
1884	59.43	18.45	20.10	0.60	1.42
1888	55.01	20.58	22.42	0.56	1.42
1892	56.52	21.71	20.86	0.17	0.75
1896	56.98	21.19	20.66	0.44	0.74
1900	56.81	19.62	22.10	0.37	1.11
1904	44.13	25.79	29.71	0.10	0.27
1908	52.96	21.18	24.81	0.39	0.65
1912	49.08	23.54	26.54	0.29	0.55
1916	51.16	22.73	25.50	0.29	0.32
1920	47.57	23.91	27.97	0.10	0.45
1924	41.29	26.13	32.02	0.06	0.48
1928	60.93	16.41	21.05	0.90	0.71
1932	49.34	26.39	23.72	0.23	0.32
1936	47.70	25.30	25.88	0.68	0.45

(continued)

Table 2.4 (continued)

Year	% Not significant	% High-high	% Low-low	% High-low	% Low-high
1940	43.35	26.94	29.16	0.32	0.23
1944	45.12	25.94	28.29	0.42	0.23
1948	57.42	20.57	20.66	0.90	0.45
1952	53.27	20.92	24.40	0.68	0.74
1956	58.83	18.44	21.47	0.84	0.42
1960	61.31	17.03	19.86	1.16	0.64
1964	68.04	18.86	12.23	0.35	0.51
1968	59.64	19.18	18.93	1.48	0.77
1972	63.69	17.15	16.99	1.38	0.77
1976	58.06	18.76	21.18	0.97	1.03
1980	58.80	19.89	19.83	0.35	1.13
1984	66.18	14.91	16.10	1.19	1.61
1988	64.39	14.90	17.65	1.23	1.84
1992	66.01	14.68	16.64	1.16	1.51
1996	65.48	15.56	16.39	1.13	1.45
2000	64.58	15.69	17.13	1.32	1.29
2004	64.30	15.44	17.85	1.16	1.25
2008	61.23	16.60	20.24	1.03	0.90
2012	60.49	17.08	20.37	1.10	0.97
2016	61.17	15.28	21.33	1.26	0.97

identified previously undiscovered features of the partisan geography of the United States.

First, and perhaps most foremost, partisan voting is not defined principally by state-level voting patterns, as the red state-blue state thesis might lead us to believe it is. On average, in American electoral history since 1828, there has been more within-state variation than between-state variation in partisan voting. Indeed, it is ironic that our current fixation on blue states and red states has taken place during a period of particularly influential within-state variation in partisan voting. As Fig. 2.1 demonstrates, within-state variation has produced approximately 60% of the variation in partisan voting since we became focused as a country on blue states and red states in 2000. This proportion is much higher than in most previous decades. In short, we have become focused on states as the units that explain the geography of voting in the U.S. at the precise time that states have become less consequential for structuring this geography in comparison to much of our electoral history in the United States.

Second, we can see that partisan voting is not randomly distributed with regard to space. Where people live is important for understanding how they vote. In each presidential election from 1828 to the present, there has been a strong positive spatial autocorrelation in county-level voting. There is, in short, a spatial structure, a political geography to voting that is important for our knowledge of voting in the United States.

Third, just as partisan voting is not randomly distributed in space, it does not randomly fluctuate from election to election. In giving a detailed accounting of the history and geography of partisan voting, a key theme that stands out is the stability of geographic voting patterns across elections. To be sure, areas of Democratic and National Republican/Whig/Republican support vary substantially across party systems, but this is with the time frame of generations. *Within* party systems, however, there is remarkable continuity, even as the candidates, issues at play, and margins of victory change from election to election. This is true even when the winning party changed. For example, from the election of Garfield (R) in 1880 to Cleveland (D) in 1884, or Truman (D) in 1948 to Eisenhower (R) in 1952, the spatial structure of partisan voting looks very similar.

The maps also help put current debate on geographic polarization in context. Spatial voting patterns in 2016, displayed in Fig. 2.50, are not a radical break from previous elections, but part of a continued, slowing evolving pattern of Democratic dominance on the coasts and Republican dominance in much of the South and Midwest. This basic pattern has characterized much of the Sixth Party System. In the next chapter, we further examine the debate on polarization and argue that, claims that we have "sorted" aside, our current era of geographic partisanship is not historically unique.

References

Abramowitz, A. I. (2010). *The disappearing center: Engaged citizens, polarization, and American democracy*. New Haven: Yale University Press.

Abramowitz, A. I., & Saunders, K. L. (2008). Is polarization a myth? *Journal of Politics, 70*(2), 542–55.

Andersen, K. (1979). *The creation of a Democratic majority, 1928–1936*. Chicago: University of Chicago Press.

Anselin, L. (1995). Local indicators of spatial association – LISA. *Geographical Analysis, 27*(2), 93–115.

Anselin, L., & Bera, A. K. (1998). Spatial dependence in linear regression models with an introduction to spatial econometrics. In A. Ullah & Giles, D. E. A. (Eds.), *Handbook of applied economic statistics* (pp. 237–289). New York: Marcel Dekker.

Ansolabehere, S., Rodden, J., & Snyder, J. M. (2006). Purple America. *The Journal of Economic Perspectives, 20*(2), 97–118.

Bafumi, J., & Shapiro, R. Y. (2009). A new partisan voter. *Journal of Politics, 71*(01), 1–24.

Bartels, L. M. (2000). Partisanship and voting behavior, 1952–1996. *American Journal of Political Science, 44*, 35–50.

Bartels, L. M. (2002). Beyond the running tally: Partisan bias in political perceptions. *Political Behavior, 24*(2), 117–50.

Basinger, S. J., & Lavine, H. (2005). Ambivalence, information, and electoral choice. *American Political Science Review, 99*(2), 169–84.

Bicknell, J. (2014). *America 1844: Religious fervor, westward expansion and the presidential election that transformed the nation*. Chicago: Chicago Review Press.

Bishop, B., with Cushing, R. G. (2008). *The Big Sort: Why the clustering of like-minded Americans is tearing us apart*. Boston: Mariner Books.

Brown, T. A. (1988). *Migration and politics: The impact of population mobility on American voting behavior*. Chapel Hill: The University of North Carolina Press.

Campbell, J. (2008). Polarization runs deep, even by yesterday's standards. In P. S. Nivola & D. W. Brady (Eds.), *Red and blue nation?* (Vol. 1, pp. 152–162). Washington, D.C.: Brookings Institution Press.

Campbell, A., Converse, P. E., Miller, W. E., & Stokes, D. E. (1960). *The American voter*. New York: Wiley.

Carsey, T. M., & Layman, G. C. (2002). Party polarization and 'conflict extension' in the American electorate. *American Journal of Political Science, 46*(4), 786–802.

Carsey, T. M., & Layman, G. C. (2006). Changing sides or changing minds? Party identification and policy preferences in the American electorate. *American Journal of Political Science, 50*(2), 464–77.

Chambers, W. N., & Burnham, W. D. (Eds.). (1967). *American party systems: Stages of political development*. Oxford: Oxford University Press.

Darmofal, D. (2015). *Spatial analysis for the social sciences*. New York: Cambridge University Press.

Darmofal, D., & Nardulli, P. F. (2010). The dynamics of critical realignments: an analysis across time and space. *Political Behavior, 32*(2), 255–283.

Dean, J. (2004). *Warren G. Harding*. New York: Henry Holt and Company.

Erikson, R. S., & Tedin, K. L. (1981). The 1928–1936 partisan realignment: The case for the conversion hypothesis. *American Political Science Review, 75*(4), 951–962.

Evans, J. (2003). Have Americans' attitudes become more polarized?—An update. *Social Science Quarterly, 84*(1), 71–90.

Fiorina, M. P. (1981). *Retrospective voting in American national elections*. New Haven: Yale University Press.

Fiorina, M. P., Abrams, S. J., & Pope, J. (2005). *Culture war? The myth of a polarized America*. New York: Pearson.

Fiorina, M. P., Abrams, S. J., & Pope, J. (2010). *Culture war? The myth of a polarized America* (3rd ed.). New York: Pearson.

Fischer, C. S., & Mattson, G. (2009). Is America fragmenting? *Annual Review of Sociology, 35*(1), 435–55.

Foner, E., & Mahoney, O. (1995). *America's reconstruction: People and politics after the civil war.* New York: HarperPerennial.

Frank, T. (2004). *What's the matter with Kansas?: How conservatives won the heart of America*. New York: Metropolitan Books.

Franklin, C. H., & Jackson, J. E. (1983). The dynamics of party identification. *American Political Science Review, 77*(4), 957–73.

Frey, W. H. (1999). Immigration and demographic balkanization: Toward one America or two? In J. W. Hughes & J. J. Seneca (Eds.), *America's demographic tapestry: Baseline for the new millennium* (pp. 78–97). New Brunswick: Rutgers University Press.

Gelman, A., Park, D., Shor, B., Bafumi, J., & Cortina, J. (2008). *Red state, blue state, rich state, poor state*. Princeton: Princeton University Press.

Gerring, J. (1998). *Party ideologies in America, 1828–1996*. Cambridge: Cambridge University Press.

Gimpel, J. G., & Schuknecht, J. E. (2003). *Patchwork nation: Sectionalism and political change in American politics*. Ann Arbor: University of Michigan Press.

Green, D. P., Palmquist, B., & Schickler, E. (2002). *Partisan hearts and minds: Political parties and the social identities of voters*. New Haven: Yale University Press.

Healy, A., & Malhotra, N. (2013). Retrospective voting reconsidered. *Annual Review of Political Science, 16* (1), 285–306.

Hetherington, M. J. (2001). Resurgent mass partisanship: The role of elite polarization. *American Political Science Review, 95*(3), 619–631.

Holt, M. F. (2003). *The rise and fall of the American whig party: Jacksonian politics and the onset of the civil war.* Oxford: Oxford University Press.
Holt, M. F. (2008). *By one vote: The disputed presidential election of 1876.* Kansas: University of Kansas Press.
Jenkins, J. A., & Sala, B. R. (1998). The spatial theory of voting and the presidential election of 1824. *American Journal of Political Science, 42*(4), 1157–1179.
Jessee, S. A. (2009). Spatial voting in the 2004 presidential election. *American Political Science Review, 103*(01), 59–81.
Johnston, R., Manley, D., & Jones, K. (2016). Spatial polarization of presidential voting in the United States, 1992–2012: The "big sort" revisited. *Annals of the American Association of Geographers, 106*(5), 1047–1062.
Jurjevich, J. R., & Plane, D. A. (2012). Voters on the move: The political effectiveness of migration and its effects on state partisan composition. *Political Geography, 31*, 429–43.
Kleppner, P. (1979). *The third electoral system, 1853–1892: Parties, voters, and political cultures.* Chapel Hill: University of North Carolina Press.
Lau, R. R., & Redlawsk, D. P. (2001). Advantages and disadvantages of cognitive heuristics in political decision making. *American Journal of Political Science, 45*(4), 951–71.
Lee, B. A. (1982). The new deal reconsidered. *The Wilson Quarterly, 6*(2), 62–76.
Leip, D. (2018) Dave Leip's Atlas of U.S. Presidential Elections. https://uselectionatlas.org/
Levendusky, M. (2009). *The partisan sort: How liberals became Democrats and conservatives became Republicans*. Chicago: University of Chicago Press.
McCarty, N., Poole, K. T., & Rosenthal, H. (2006). *Polarized America: The dance of ideology and riches*. Cambridge: MIT Press.
McGhee, E., & Krimm, D. (2009). Party registration and the geography of party polarization. *Polity, 41*(3), 345–67.
McKee, S. C., & Teigen, J. M. (2009). Probing the reds and blues: Sectionalism and voter location in the 2000 and 2004 U.S. presidential elections. *Political Geography, 28*(8), 484–95.
Morrill, R., Knopp, L., & Brown, M. (2007). Anomalies in red and blue: Exceptionalism in American electoral geography. *Political Geography, 26*(5), 525–53.
Nardulli, P. F. (1995). The concept of a critical realignment, electoral behavior, and political change. *American Political Science Review, 89*(1), 10–22.
Nie, N., Verba, S., & Petrocik, J. (1979). *The changing American voter*. Cambridge: Harvard University Press.
Nivola, P. S., & Galston, W. A. (2008). Toward depolarization. In P. S. Nivola, & D. W. Brady (Eds.), *Red and blue nation?* (Vol. 2, pp. 235–84). Washington, D.C.: Brookings Institution Press.
Sanders, E. (1999). *Roots of reform: Farmers, workers, and the American state* (pp. 1877–1917). Chicago: University of Chicago Press.
Schattschneider, E. E. (1960). *The semisovereign people: A realist's view of democracy in America.* New York: Holt, Rinehart and Winston
Schlesinger, A. M., Jr. (1945). *The age of Jackson*. Boston: Little, Brown and Company.
Shaw, D. (2012). If everyone votes their party, why do presidential election outcomes vary so much? *The Forum, 10*, 1.
Stoker, L., & Jennings, M. K. (2008). Of time and the development of partisan polarization. *American Journal of Political Science, 52*(3), 619–35.
Summers, M. W. (2000). *Rum, romanism, and rebellion: The making of a president, 1884.* Chapel Hill: University of North Carolina Press.
Theriault, S. M. (2008). *Party polarization in Congress*. Cambridge: Cambridge University Press.

Chapter 3
Geographic Polarization in Historical Perspective

In the days before the inauguration of Donald Trump, the PBS show *Frontline* aired a two-part documentary called "Divided States of America." The show starts with a famous quote from Obama from his address at the 2004 Democratic National Convention: "There is not a liberal America and a conservative America, there is the United States of America." The documentary then catalogs how, instead of creating a unified country, the division between "liberal" and "conservative" America only exacerbated since the election of Obama in 2008. This deepening division and animosity, the documentary asserts, allowed for the election of "the ultimate outsider," Trump. The opening montage sums it up with a commentator stating "Obama came in with the mandate of changing Washington. By his very presence, he forced more polarization and gridlock than we had seen in the eight years prior."

The argument that there is a liberal and conservative, or a red and blue, America, and that the division between these two camps has only grown in the past few elections, is not novel. But is it accurate? And, given the scope of U.S. history, is it historically atypical? Using Bill Bishop's popular and influential book *The Big Sort* as a foil, we argue that this assessment is only partly accurate and not historically atypical. Bishop asserts that, because of the increased propensity to migrate based on lifestyle choices, the country is increasingly "clustered" in a way that "tears us apart." This chapter, however, reviews a litany of literature that suggests, while it is true that the country has geographically sorted by party over recent years, it is likely not due to individuals' migration decisions.

The chapter then uses county level election and demographic data from 1828 to 2016 to do what previous research has not adequately done—put the argument for geographic polarization in historical context. We find here that, while geographic polarization may be on the rise when one uses 1976 as a baseline (as Bishop does), our era is not particularly divided given a fuller scope of U.S. history. Finally, the chapter analyzes whether the connection between county educational levels and voting patterns—a key mechanism for The Big Sort's argument—has grown over

D. Darmofal, R. Strickler, *Demography, Politics, and Partisan Polarization in the United States, 1828–2016*, Spatial Demography Book Series 2,
https://doi.org/10.1007/978-3-030-04001-7_3

the past four decades. While this analysis is supportive of the "sorting" thesis, the strongest relationships between education and partisan voting have manifested only in the past few elections. Time is needed to tell whether this is a historical blip or an enduring pattern.

3.1 Bishop's Argument

After the 2000 election, there was much discussion of geographic polarization, or the increasing geographic division of the populace into homogenous enclaves based on partisanship, voting patterns, issue positions, or other dimensions. Commentators focused on the irreconcilability of "red states" representing "an entire geography (rectangular borders in the country's midsection), an iconography (Bush in a cowboy hat), and a series of cultural cliches (churches and NASCAR)" and "blue states" representing "something on, and of, the coastal extremes, urban and latte-drinking" (Farhi 2004). In response, political geographers and public opinion scholars have contended that polarization is not as strong as the popular imagination suggests (Darmofal and Strickler 2016). The scholarly consensus was that localities on the political map were not less competitive than in the past (Klinkner 2004), and any "red versus blue" divergence in vote patterns was not mirrored by a divergence in attitudes or ideological outlook (Fiorina et al. 2010; Ansolabehere et al. 2006).

With *The Big Sort*, we seem to be repeating this cycle—a bold popular claim concerning geographic polarization that is now being tempered by academic research. In the book, Bishop contends that we have become so politically segregated that "ways of life now have a distinct politics and a distinct geography" (14). The empirical backbone of *The Big Sort* is the rise of "landslide" counties since 1976. These are counties where the margin of victory for one presidential candidate over another is 20 percentage points or more. From 1976 to 2004, Bishop notes that U.S. elections have gotten more competitive overall, with the 2000 and 2004 elections coming down to differences in just a small fraction of the electorate in key swing states. However, he shows that localities have grown more electorally lopsided over that time, with the percentage of voters living in landslide counties growing from less than a quarter in 1976 to nearly half in 2004. He argues against the notion that polarized voting patterns belie a moderate public forced to choose between two ideologically extreme candidates (as Fiorina et al. 2010, contend). He also argues that this trend is not driven solely by successful conservative activism and an asymmetrical polarization of the Republican party. Instead, Bishop contends that the growth in landslide counties represents:

> "(B)alkanized communities whose inhabitants find other Americans to be culturally incomprehensible; a growing intolerance for political differences that has made national consensus impossible; and politics so polarized that Congress is stymied and elections are no longer just contests over policies, but bitter choices between ways of life" (14).

Bishop argues that these trends are fueled by increasing economic prosperity, an increasingly footloose public, and the increasing relevance of culture to partisan politics. Drawing on the "creative class" thesis of Florida (2002), he contends that the college educated and professional class base migration decisions less on traditional motivators such as employment opportunity, family, or community, and more on the lifestyle and cultural amenities they prefer. This creates an "unconscious decision to cluster in communities of like mindedness" (15), as well as increased homogeneity for those not in the educated class that are not able to move. At the same time, since the 1960s politics has taken a post-materialist turn, where social and religious issues have since increasingly become wrapped into partisan conflict. Thus, Bishop argues one's party, one's lifestyle, and one's decisions on where to live are all cut from the same cloth. While he does not argue that people move for explicitly partisan reasons, he contends that geographic partisan polarization results from the public's migration decisions.

To bolster his claim of a 'sorted' America, in addition to the growth in landslide counties since 1976, Bishop provides three tests (Chapter Three in his book). For the first test, he shows that roughly half of 2004 voters lived in a county that had not changed from red to blue, or vice versa, since 1980. These counties, moreover, tend to have increasing partisan margins of victory with each subsequent election. For the second, he shows that differences in demographic, socioeconomic, and key opinion measures are significantly different between habitants of modern Republican and Democratic counties. For the third test, he shows that the divide between landslide Republican and Democratic counties for rates of educational attainment, rates of church attendance, and percentages of non-white and foreign born citizens has grown since the 1970s.

The political consequences of a sorted America are dire, according to Bishop. Gridlock is constant, and partisans are unlikely to agree on a set of shared facts, much less a policy. Policymaking thus primarily occurs through unelected bureaucracy, or through ideologically charged initiatives adopted in politically homogenous states or localities. Members of the House of Representatives are firmly entrenched in safe districts and senators in safe Senate seats. Lawmakers are more likely to listen to ideologically-driven donors and activists, as moderates are squeezed out of both parties. In short, the "big sort" is a threat to our democracy.

3.2 Recent Assessment of *The Big Sort*

Bishop offers a provocative, yet intuitive hypothesis for the growth of partisan conflict over the past few decades. But the empirical evidence he cites for his claim that the U.S. is politically "sorted" due to in-migration is somewhat thin (Darmofal and Strickler 2016). As such, since the publication of his book political scientists have put his "sorting" thesis to the test, with mixed results. Some push back on the notion that the country is sorted at all, or at least in the way Bishop argues. For instance, Abrams and Fiorina (2012) argue that presidential election results can exaggerate the extent to which counties are polarizing. For example, one county's voters can be on the whole only slightly conservative, but given a choice between a very liberal Democratic and a very conservative Republican candidate, a "landslide" majority may prefer the Republican candidate. They, as well as McGhee and Krimm (2009), instead examine geographic patterns for party *registration* data, which they argue is a more reliable barometer for political preference. Instead of finding a sorting over recent decades, they instead find an increase in independent registration. In a similar vein, Strickler (2016) contends that presidential vote choice can be motivated by a wide range of factors beyond cultural and political values. Moreover, in line with Levendusky and Pope (2011), mean differences in opinion or demographic indicators between "red" and "blue" counties can be misleading, because averages may belie a great deal of variance and therefore overlap in opinion. Examining 2012 election and survey data, he shows that, for political values across economic, social, and cultural dimensions, there is far more common ground than division even between "landslide" red and "landslide" blue counties.

Others, however, agree with Bishop that, at least in recent years, geographic partisan segregation is on the rise. It is true that, for many decades, Democratic voters have been sorted into urban areas; concentration of left leaning parties in urban areas is a constant across industrialized countries (Rodden 2010). In fact, using 2000 Presidential election data from all 50 states, Chen and Rodden show that, because of partisan sorting, homogeneous concentration of Democratic voters in urban Congressional districts is impossible to avoid with *any* effort to draw contiguous districts, gerrymandering or not (2013). Massey Rothwell, and Domina, moreover, argue (like Bishop) that political geographic segregation at the micro-level (the neighborhood or census tract) has increased over the last third of the twentieth century. Drawing on both census data and General Social Survey data of political attitudes, they demonstrate that this is largely driven by increasing geographic segregation based on income class (2009). Likewise, Johnston, Manley, and Jones find that, from 1992 to 2012, geographic segregation of partisan voting has increased when examining segregation at various levels. Census regions within the nation, states within regions, and counties within states have all become more segregated (2016). Focusing on specific states and localities, still more scholars have found increasing geographic polarization at smaller geographic levels, such as the precinct or block level (Kinsella et al. 2015; Sussell 2013; Myers 2013).

Some critics, though, while acknowledging some increased sorting, question the role of migration in this process. In addition to migration, political sorting could also occur through conversion of voters from one party to another, or by generational replacement of voters within communities. Recent surveys suggest partisans have both an explicit preference for living in communities with like partisans, as well as differences in desired neighborhood amenities that could also fuel migration-based partisan sorting (Political Polarization in the American Public 2014). This has been backed up by experimental research showing that partisans are less likely to view properties favorably if they are in an "outparty" neighborhood (Gimpel and Hui 2015), and are less likely to express sense of belonging to a politically incongruous neighborhood (Motyl et al. 2014). Mummolo and Nall (2017), however, contend that findings like these can be misleading because they discount more important factors for moving decisions. Using a conjoint experimental approach on a nationally reflective sample of 4,800 self-identified Democrats and Republicans, they show that, even for strong partisans, things like crime, schools, and commuting time matter more than politics in a move choice. They then show that these partisan individuals, in their actual last move, did not move to politically homogenous zip codes. In fact, both Democrats and Republicans displayed a slight tendency to move to Republican communities. This is a finding backed up by Martin and Webster (2017), who draw on public voter files to demonstrate that, while there is a statistically significant tendency for partisans to move to politically like communities, this tendency is very small (see also Cho et al. 2013). They then use simulations to demonstrate that migration alone only makes a small contribution to the observed level of partisan geographic sorting, and other factors (such as replacement or conversion) must play a primary role.

Some argue that race plays a more consequential role in geographic sorting. While Bishop focuses on culture and lifestyle conflict as drivers, there is surprisingly little discussion of race as a component of, or primary driver of, this conflict. This is despite the long-standing role race has played in partisan politics throughout history, from the rise of the Republican Party precipitating the Civil War to the dominance of the "solid" Democratic South, due to racial disenfranchisement, through most of the twentieth century. In this century, the Democratic and Republican party coalitions have increasingly segregated by race. Since the election of Obama in 2008, whites (particularly whites with racially conservative attitudes) have increasingly identified with the Republican Party, while blacks and other non-white groups have identified strongly with the Democratic Party (Tesler 2016). At the same time, while the country has grown less racially segregated over the past couple of decades, racial segregation by state, county, and census tract is still higher than segregation based on income (Massey et al. 2009). Thus, with race increasing as a driver of partisan politics, it stands to reason that racial segregation would produce increasing geographic polarization.

Many argue, moreover, that racial factors with regard to *migration* further fuel partisan geographic sorting. There is a well-established literature pointing to racial factors contributing to migration patterns; consider, for example, the "Great Migration" of African-Americans to Northern communities (Schelling 1971) or the prevalence of "white flight" from inner cities during the twentieth century (Duncan and Duncan 1957). Moreover, Scher posits that an influx of middle class whites to the South through the middle of the twentieth century was one key contributing factor to the region's subsequent Republican transformation (1997). Anastasopoulos (2017), furthermore, advances a Migration-Sorting theory, whereby changes in neighborhood racial composition produce migration decisions based on one's tolerance for diversity. Those more open to diversity are both more likely to remain when neighborhoods become more racially mixed and more likely to vote Democratic. Those with more closed personality are more likely to both leave and vote Republican. Thus, exogenous changes in neighborhood racial demographics produce partisan sorting. The author finds evidence of this theory in Houston, where migration from African-Americans after Hurricane Katrina in 2005 produced Republican flight. With this study and others in mind, it is hard to ignore race as a key factor in both partisan conflict and patterns of in-migration in the United States. Again, Bishop pays too little attention to this factor.

In sum, the recent research suggests that, notwithstanding the rise in independent identification and common ground on issues Democrats and Republicans share, Bishop is correct that the country has "sorted" along party voting lines in recent decades. More suspect is his argument that migration, as opposed to other factors such as conversion or voter replacement, is driving the sorting. This research, however, by and large relies on experimental evidence, or data such as surveys or voter files that are only available for recent years. Most of this research does not take an historical perspective or examine sorting prior to WWII. Bishop himself acknowledges that this is a potential limitation, as he states that "some argue that, viewed over the centuries, the increase in geographic segregation since the mid-1970s has been minor, a subtle fluctuation, and compared to the Civil War period, that is certainly the case" (25). However, the bulk of his data uses 1976 as a baseline. As we will show, this is an atypical year of partisan detente. Extending the timeline of analysis serves to temper claims that we are tearing apart in an historically unusual way.

3.3 Historical Examination of Geographic Polarization

Bishop's conclusion that geographic polarization is becoming a concern rests, of course, on his choice of 1976 as a baseline. If 1976 is a representative baseline, we might have reason to agree with Bishop that there has been a dramatic (and potentially problematic) increase in geographic polarization over time. If, instead, 1976 is an unrepresentative baseline we should be less sanguine about accepting Bishop's conclusions. What does geographic polarization look like when placed in a longer historical perspective? To examine this, we replicate Bishop's approach, but extend the analysis back in time to the beginning of mass voter participation in the United States with the advent of Jacksonian democracy in 1828. We also bring the analysis up to the most recent presidential election of 2016. Specifically, as Bishop did, we identify the percentage of voters voting for the Democratic or Republican Party who lived in "landslide" counties in which either the Democratic or Republican candidate won the county by a margin of 20 percentage points or more.

Figure 3.1 presents the percentage of two-party voters living in "landslide" counties in each presidential election from 1828 through 2016. Figure 3.2 breaks out voters from these landslide counties by their partisanship. What stands out is how unrepresentative 1976 is as a baseline election. In fact, the 1976 presidential election was one of the least polarized presidential elections since the beginning of mass voting in 1828. And it remains one of the least polarized elections. Only 26.59% of voters lived in landslide counties in 1976. The only other presidential elections with fewer than 30% of voters living in landslide counties were the 1852 and 1916 elections. The mean percentage of voters living in landslide counties prior to 1976 was, in fact, nearly 20 points higher than in 1976, at 44.85%. Figure 3.2

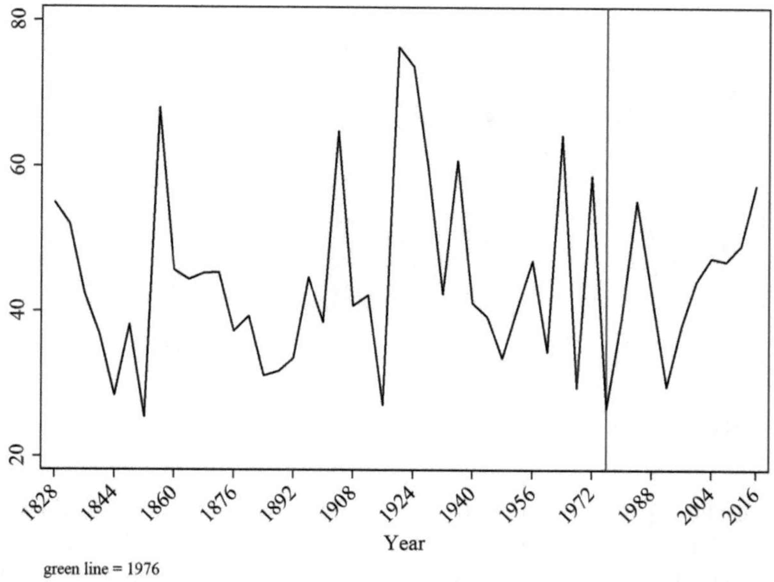

Fig. 3.1 Percentage of two-party voters living in landslide counties

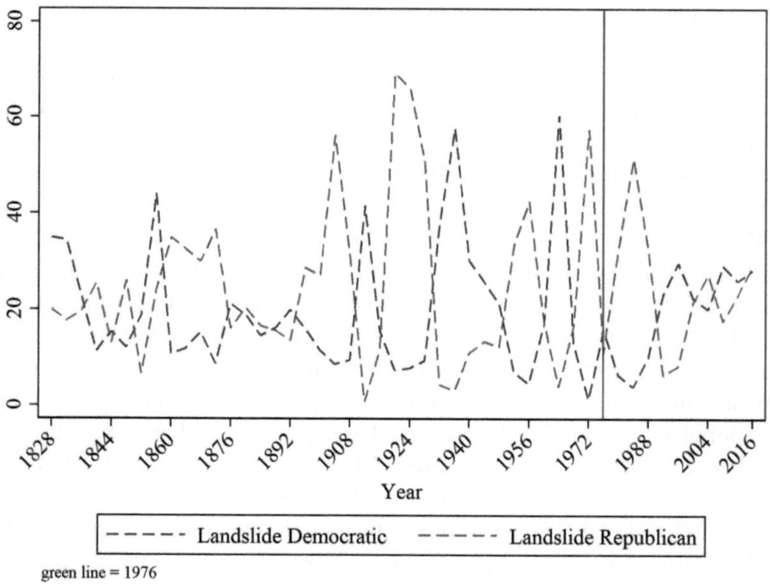

Fig. 3.2 Percentage of two-party voters living in landslide counties, by party

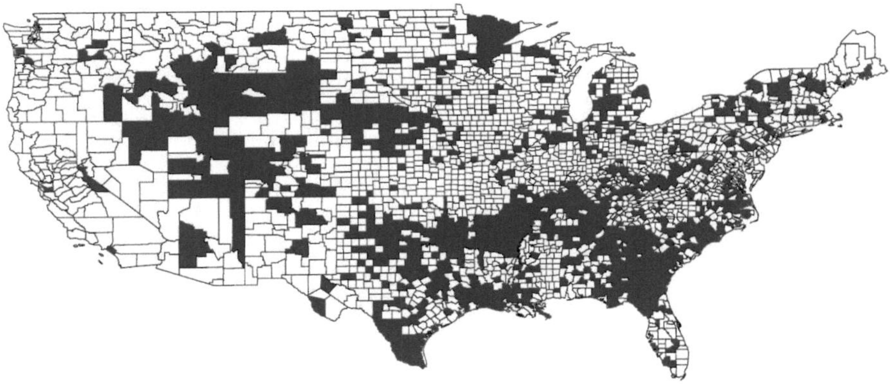

Fig. 3.3 Map of landslide counties, 1976

illustrates how partisan history has oscillated between Democratic dominance, National Republican/Whig/Republican dominance, and times where both parties had significant numbers of "landslide" voters. 1976, though, represents a nadir for *both* parties with regard to landslide voters. In short, any analysis of polarization based on a 1976 baseline rests on very shaky ground.

Have elections after 1976 been marked by unusually high polarization, as Bishop argues? No, they haven't. The mean percentage of voters living in landslide counties after 1976 is nearly the same as prior to 1976, at 44.95%. The elections of Barack Obama in 2008 and 2012 have often been cited as a high-water mark for the partisan geographic sorting Bishop describes, with Democrats inhabiting urban, coastal areas and Republicans inhabiting rural areas in the middle of the country. While the urban-rural divide is very real (Cramer 2016), using the landslide metric, it is historically common. In 2008, 47.6% of major party voters lived in landslide counties, and in 2012 49.8% of major party voters lived in such areas. This is only a few percentage points higher than the historical average prior to 1976. There is an uptick in the proportion of two-party voters living in landslide counties for the 2016 election, with 57.6% of voters living in such areas. However, this is still not a complete anomaly, as nine of 48 Presidential elections since 1828 had landslide percentages higher than this. What's more, the 2016 presidential election was very unique, with two historically unpopular candidates following an Obama presidency that fueled strong racial divides between the parties (Tesler 2016). Thus, it is not immediately clear whether this election is a one-off or will be part of a long-term trend. It is too early to tell.

Figures 3.3, 3.4, 3.5, and 3.6 further illustrate the historically atypical nature of the 1976 election Bishop uses as a baseline. These figures include maps similar to what one finds in *The Big Sort*, displaying the geographic layout of "sorted" versus "competitive" Democratic and Republican counties in the United States. Figures 3.3 and 3.4 display maps for both the 1976 and the 2016 election. Areas in blue denote Democratic "landslide" counties, areas in red denote Republican "landslide" counties, and white areas denote non-landslide (or competitive) counties. Looking at these maps alone, it appears that geographic polarization is on the rise, as there

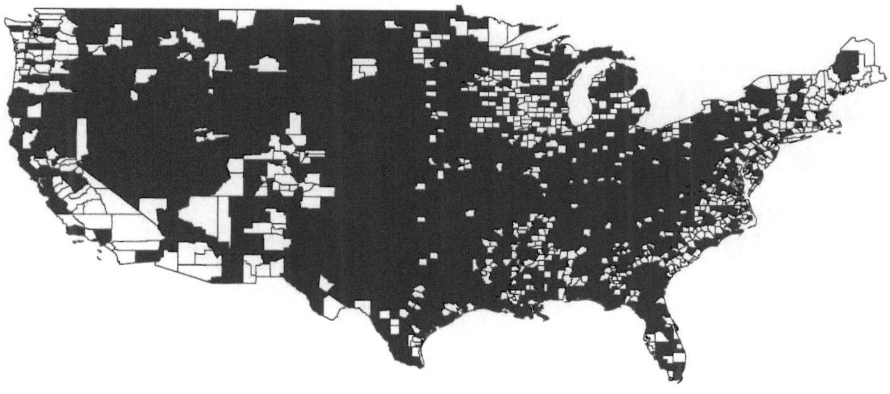

Fig. 3.4 Map of landslide counties, 2016

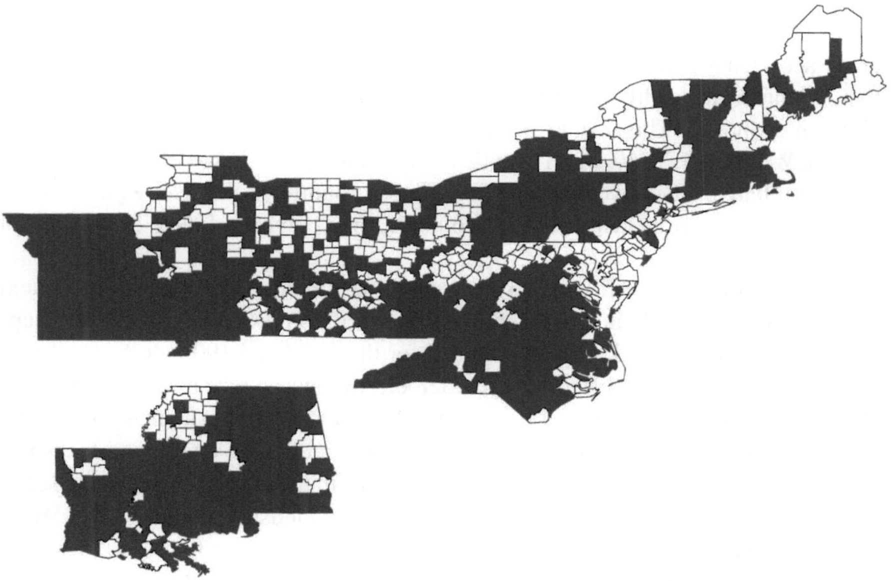

Fig. 3.5 Map of landslide counties, 1828

are clearly more "sorted" areas of the country in 2016 than 40 years prior. Figure 3.3 reflects the fact that Jimmy Carter, a Georgia Democrat, did very well in the South in 1976. In fact, no Democratic Presidential candidate has received a majority of votes in Southern states since. Gerald Ford, in contrast, fared better in the Midwest, the Rocky Mountain states, and parts of the Northeast. Beyond this, both a majority of counties (62.15%) were competitive, and a majority of two-party voters (73.41%) lived in counties that were competitive.

The growing urban/rural divide in voting patterns (Chen and Rodden 2013) is reflected in Fig. 3.4. It is awash in red, illustrating Trump's dominance in rural

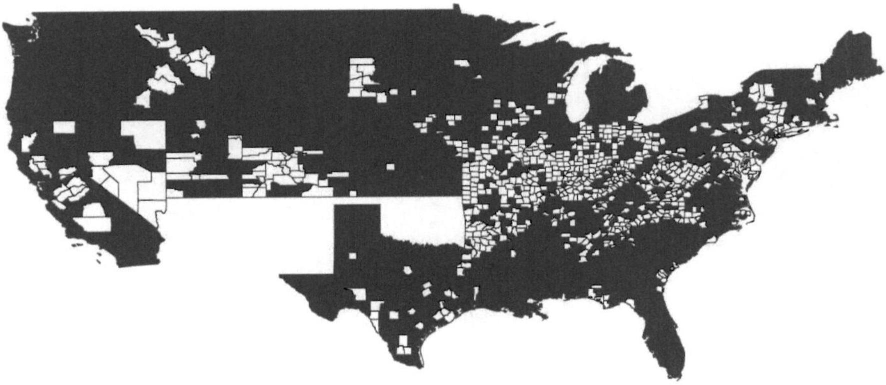

Fig. 3.6 Map of landslide counties, 1904

areas across the country in 2016. In fact, a casual look at the map may suggest that Trump won handily, with Clinton only winning areas along the Eastern and Western seaboard. This would belie the dominance Clinton had in urban areas, though. While only 7.34% of all counties were "landslide" Clinton counties, 28.27% of all two-party voters lived in landslide Clinton counties. Likewise, while 70.65% of all counties were landslide Trump counties, they comprised only 29.18% of all two party voters.

What's more, Figs. 3.5 and 3.6 provide some much-needed historical context to this alleged growth in geographic polarization. As examples, they are maps showing the location of landslide counties for the 1828 and 1904 elections. These two elections were very different. The 1828 election was a race between Democrat and eventual winner Andrew Jackson and National Republican incumbent John Quincy Adams. It represents the advent of the Second Party System, and Jackson's Democratic Party was the first to mobilize partisans in the electorate on a mass scale. In this election, 54.98% of two-party voters lived in landslide counties. 34.95% of two party voters lived in landslide Democratic counties. Figure 3.5 shows that these counties are concentrated in the South, what at the time was the Western part of the country, and Pennsylvania. 20.03% of two party voters lived in landslide National Republican counties. Figure 3.5 shows that these counties are concentrated in the Northeast, particularly Massachusetts, Connecticut, Rhode Island, and Vermont.

One finds a higher proportion of voters living in landslide counties in 1904. After taking office following McKinley's assassination, the 1904 election was Theodore Roosevelt's first election. He won handily, 64.5% of two-party voters living in landslide counties for this election, with 55.6% living in landslide counties for Roosevelt, while only 8.9% lived in landslide counties for his Democratic opponent, Alton B. Parker. However, Fig. 3.6 shows that, despite the runaway victory, the "system of 1896" produced a sharp partisan divide between the North and South. While only comprising less than 9% of two-party voters, 31.62% of all counties in the country were Parker "landslide" counties. These counties are almost exclusively concentrated in the South. In contrast, 43.39% of counties in the country were Roosevelt "landslide" counties, almost exclusively concentrated in the North, Midwest, and West. Overall, Figs. 3.5 and 3.6 demonstrate that 1976 is a poor baseline to assess geographic polarization. Yes, there may be a stronger inter-county partisan divide in 2016 as compared to 1976 (and even 2012). However, it is not historically unique. And this analysis highlights the importance of placing contemporary elections in an historical context.

3.4 Education and the Geography of Presidential Voting

Examining voting patterns, however, is only part of the story. What makes Bishop's work novel is not only that he claims we are polarizing geographically, but that this is driven by "a post-materialist Tiebout migration based on non-economic goods, as people have sought out places that best fit their ways of life, their values, and their politics," (199). Educational attainment is a key part of this argument. College-educated individuals, with greater economic opportunity and less commitment to home communities than in the past, increasingly seek out urban centers of culture. The areas these individuals migrate to are increasingly homogenous, adopting left-leaning and Democratic politics, while the areas they leave behind are increasingly becoming more homogenous, adopting right-leaning and Republican politics. This argument is influenced by the 'creative class' thesis of Richard Florida. Florida (2002) suggests that a fruitful strategy for local economic development is for cities to develop support for professional growth and technological innovation, as well as a strong urban culture where citizens engage in myriad experiential and creative public activities. While some have critiqued the viability of this development strategy (Peck 2005), the creative class thesis is still influential among urban planners and local government officials today (Florida 2014).

If Bishop's argument is correct, we should see over the past few decades a strengthening relationship between the education level of a county's citizens and their partisan vote preference. And indeed, this is what the data show. For all presidential elections since 1976, Table 3.1 displays the correlation between the Democratic margin of the two-party vote and the proportion of citizens over 25 with at least a completed bachelor's degree, at the county level. The table indicates that the connection between aggregate education and partisanship has flipped over the past four decades. In 1976, county education level was moderately and negatively correlated with the Democratic two-party margin of victory. Since then, the correlation has steadily increased, to the point where in 2016 the correlation was positive and higher, at an absolute level, than in 1976.

Figure 3.7 further demonstrates this trend. This figure shows the mean county-level percentage of all citizens over 25 with a least a bachelor's degree, for both landslide Democratic counties and landslide Republican counties, in each election since 1976. To be sure, the figure partly reflects a secular trend of higher educational attainment across the country. In 1976, the average percentage of those over 25 with

Table 3.1 Correlation between county-level education and democratic vote share

Year	Pearson's r
1976	−0.35
1980	−0.22
1984	−0.10
1988	−0.07
1992	−0.08
1996	−0.06
2000	0.09
2004	0.21
2008	0.31
2012	0.30
2016	0.46

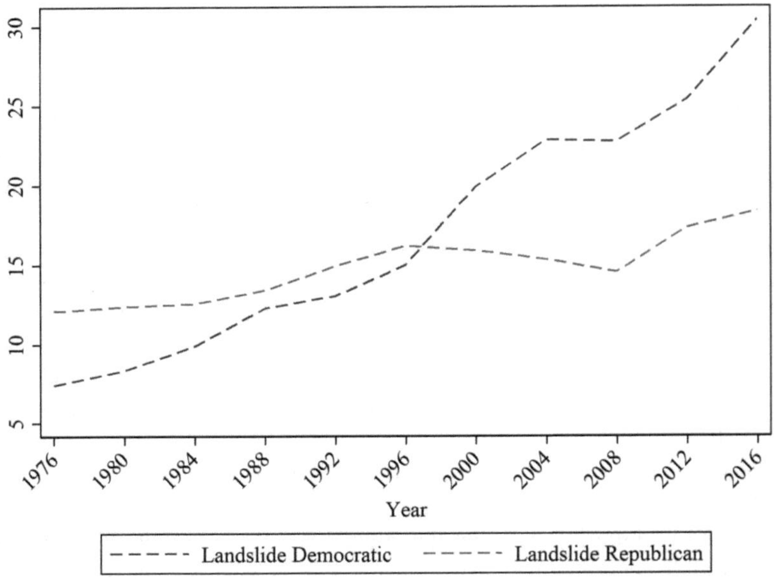

Fig. 3.7 Percent over 25 with bachelor's degree, by party voting

a bachelor's degree for all counties was 9.78%; in 2016, this percentage was 20.76%. What is striking, though, is the differing trajectories between landslide Democratic and Republican counties. The mean level of education in landslide Republican counties has only slightly increased. In 1976, the percentage of citizens over 25 with a bachelor's degree in these counties was 12.06%, and in 2016, the percentage for these counties was 18.34%. The educational level for landslide Democratic counties, though, has increased dramatically. In 1976, the mean percentage of citizens with at least a bachelor's degree for these counties was only 7.41%; in 2016, the percentage was 30.36%.

If anything, Bishop may be prescient on this point. In *The Big Sort*, his analysis stops at 2004. However, the positive relationship between county-level education and Democratic vote share is stronger in recent elections. The correlation between mean education and Democratic two-party margin of victory jumps from 0.21 in 2004 to 0.46 in 2016. Moreover, the mean percentage of individuals over 25 with a bachelor's degree in landslide Democratic counties jumps from 15.32% in 2004 to 30.36% in 2016. This may give some support to Bishop's argument that the economic opportunity afforded through education increasingly allows people to "sort" based on lifestyle choices, causing partisan geographic polarization. However, it is important to reiterate a few things discussed earlier. First, as with Bishop's work, this data is aggregated at the county level, and the effect of education on partisan choice is better studied at the individual level. Second, this data does not suggest whether migration, as opposed to conversion or replacement, is driving the county-level relationship between education and partisanship. Finally, this trend is recent, manifesting most strongly in only the past few elections. Time will tell if it continues over the course of multiple decades.

3.5 Conclusion

Is geographic polarization real? The answer to that question depends on the time frame one is considering. As Bishop points out, the U.S. political map, at the county level and at least with looking at election results, has gotten more segregated. The proportion of the voting public living in "landslide" Democratic or Republican counties has more than doubled from 1976 to 2016. In the last election, Democratic voters were strongly concentrated in New England and the coasts, while Republican voters were concentrated in the South and throughout "flyover country." Moreover, the connection between partisanship and education at the county level, a key mechanism for Bishop's "sorting" hypothesis, has gotten stronger since 1976. While a county's education level was modestly and negatively correlated with Democratic vote share in 1976, 2016 saw a correlation that was positive and stronger at an absolute level.

Is this recent polarization driven by a "post-materialist Tiebout style migration," as Bishop contends? Our analysis cannot answer whether that is the case, or whether conversion or generational replacement are driving these phenomena. Previous research, however, does cast doubt on the "sorting" thesis. While politics can play a role, as an explicit factor it only plays a minor role in individuals' migration choices (Cho et al. 2013). Even choices in "post-material" lifestyle and cultural amenities, which can (as Bishop argues) indirectly lead to geographic polarization, are less strong factors than material concerns such as jobs, crime, schools, etc. (Mummolo and Nall 2017). Bishop's "sorting" thesis, moreover, does not consider the role that race plays in migration decisions specifically (Anastasopoulos 2017) or partisan division more generally (Key 1949; Tesler 2016).

Is recent geographic polarization a historically unique phenomenon? Our evidence suggests not. In his analysis, Bishop uses a baseline that is anomalous in its lack of polarization. Thus, while indeed there has been more "sorting" over the past four decades, it is well within a normal range when looking at a longer time frame. Even the 2016 election, where over half of two-party voters lived in landslide Democratic or Republican counties, has historical precedent in the Jacksonian Era, the Civil War era, and the turn of the twentieth century. Furthermore, we see the strongest levels of geographic polarization, and the strongest relationship between education and partisanship, only with the 2016 election. While Chapter Five discusses the parallels between the 1828 and 2016 elections, in many ways the candidates and issues at play were, to say the least, unique. It is thus unclear where these trends in partisan voting will continue in the future. Indeed, leading up to the 2018 midterm and 2020 Presidential elections, both parties are engaging in an internal debate as to what it means to be "Democrat" or "Republican." As such, we may see a disruption of the polarization trend as party coalitions shift. Time, and more elections, will tell whether recent geographic polarization is a lasting phenomenon.

Appendix

Table 3.2 Percentage of voters living in "Landslide" counties, 1828–2016

Year	% landslide	% dem. landslide	% rep. landslide
1828	54.98	34.95	20.03
1832	52.02	34.34	17.68
1836	42.43	22.79	19.64
1840	36.81	11.31	25.50
1844	28.40	15.34	13.06
1848	38.12	12.07	26.05
1852	25.39	18.81	6.58
1856	68.02	43.96	24.06
1860	45.68	10.76	34.92
1864	44.34	11.85	32.49
1868	45.21	15.26	29.95
1872	45.37	8.74	36.63
1876	37.20	21.15	16.05
1880	39.33	18.92	20.41
1884	31.14	14.55	16.59
1888	31.77	16.17	15.60
1892	33.52	19.95	13.57
1896	44.66	15.96	28.70
1900	38.47	11.42	27.05
1904	64.80	8.63	56.17

(continued)

Table 3.2 (continued)

Year	% landslide	% dem. landslide	% rep. landslide
1908	40.75	9.52	31.23
1912	42.24	41.45	0.79
1916	27.09	15.72	11.37
1920	76.40	7.33	69.07
1924	73.70	7.80	65.90
1928	59.97	9.29	50.68
1932	42.37	38.04	4.33
1936	60.81	57.68	3.13
1940	41.17	30.21	10.96
1944	39.24	25.85	13.39
1948	33.53	21.26	12.27
1952	40.40	7.00	33.40
1956	46.99	4.56	42.43
1960	34.44	16.84	17.60
1964	64.30	60.36	3.94
1968	29.48	12.88	16.60
1972	58.75	1.32	57.43
1976	26.59	15.79	10.80
1980	38.95	6.41	32.54
1984	55.32	3.92	51.40
1988	42.51	10.04	32.47
1992	29.65	23.32	6.33
1996	38.04	29.76	8.28
2000	44.11	22.54	21.57
2004	47.41	20.09	27.32
2008	46.89	29.25	17.64
2012	49.16	26.05	23.11
2016	57.45	28.27	29.18

Table 3.3 "Landslide" counties, 1828–2016

Year	% landslide	% dem. landslide	% rep. landslide
1828	68.09	55.82	12.27
1832	72.27	63.42	8.85
1836	57.53	34.47	23.06
1840	50.54	21.27	29.27
1844	45.28	29.99	15.29
1848	47.80	28.51	19.29
1852	51.32	43.46	7.86
1856	84.39	73.32	11.07
1860	34.33	16.27	18.06
1864	60.91	16.05	44.86

(continued)

Table 3.3 (continued)

Year	% landslide	% dem. landslide	% rep. landslide
1868	60.80	23.97	36.83
1872	59.97	24.16	35.81
1876	58.80	38.32	20.48
1880	57.79	35.43	22.36
1884	50.66	30.89	19.77
1888	50.50	31.97	18.53
1892	56.57	33.43	23.14
1896	54.20	36.59	17.61
1900	52.78	30.45	22.33
1904	75.01	31.62	43.39
1908	55.47	30.43	25.04
1912	58.05	56.41	1.64
1916	50.83	41.84	8.99
1920	75.05	28.45	46.60
1924	75.86	33.05	42.81
1928	64.86	17.32	47.54
1932	68.40	64.92	3.48
1936	65.08	60.80	4.28
1940	59.70	43.90	15.80
1944	60.42	39.11	21.31
1948	47.34	32.70	14.64
1952	63.95	18.54	45.41
1956	59.03	16.77	42.26
1960	47.63	18.57	29.06
1964	54.78	42.87	11.91
1968	37.95	7.53	30.42
1972	79.63	0.39	79.24
1976	37.85	25.30	12.55
1980	50.69	9.85	40.84
1984	65.53	2.12	63.41
1988	44.81	5.55	39.26
1992	26.36	13.29	13.07
1996	32.70	15.19	17.51
2000	52.04	6.18	45.86
2004	59.80	5.31	54.49
2008	55.33	10.43	44.90
2012	64.13	8.70	55.43
2016	77.99	7.34	70.65

References

Abrams, S. J., & Fiorina, M. P. (2012). The big sort that wasn't: A skeptical reexamination. *PS: Political Science and Politics, 45*(2), 203–210.

Anastasopoulos, L. J. (2017). Migration, immigration, and the political geography of American cities. *American Politics Research*. Online First Edition: 1–18.

Ansolabehere, S., Rodden, J., & Snyder, J.M. (2006). Purple America. *The Journal of Economic Perspectives, 20*(2), 97–118.

Bishop, B., with Cushing, R. G. (2008). *The Big Sort: Why the clustering of like-minded Americans is tearing us apart*. Boston: Mariner Books.

Chen, J., & Rodden, J. (2013). Unintentional Gerrymandering: Political geography and electoral bias in legislatures. *Quarterly Journal of Political Science, 8*(3), 239–69.

Cho, W. K. T., Gimpel, J. G., & Hui, I. S. (2013). Voter migration and the geographic sorting of the American electorate. *Annals of the Association of American Geographers, 103*(4), 856–870.

Cramer, K. J. (2016). *The politics of resentment: Rural consciousness in Wisconsin and the rise of Scott Walker*. Chicago: University of Chicago Press.

Darmofal, D., & Strickler, R. (2016). Bringing together spatial demography and political science: Reexamining the big sort. In F. M. Howell & J. R. Porter (Eds.), *Recapturing space: New middle-range theory in spatial demography* (pp. 139–156). Dordrecht: Springer.

Duncan, O. D., & Duncan, B.. (1957). *The Negro population of Chicago: A study of residential segregation*. Chicago: University of Chicago Press.

Farhi, P. (2004). Elephants are red, donkeys are blue, color is sweet, so their states we hue. Washington Post, November 2, 2004. http://www.washingtonpost.com/wp-dyn/articles/A17079-2004Nov1.html

Fiorina, M. P., Abrams, S. J., & Pope, J. (2010). *Culture war? The myth of a polarized America* (3rd ed.). New York: Pearson.

Florida, R. (2002). *The rise of the creative class*. New York: Basic Books.

Florida, R. (2014). The creative class and economic development. *Economic Development Quarterly, 28*(3), 196–205.

Gimpel, J. G., & Hui, I. S. (2015). Seeking politically compatible neighbors? The role of neighborhood partisan composition in residential sorting. *Political Geography, 48*, 130–42.

Johnston, R., Manley, D., & Jones, K. (2016). Spatial polarization of presidential voting in the United States, 1992–2012: The 'Big Sort' revisited. *Annals of the American Association of Geographers, 106*(5), 1047–62.

Key, V. O., Jr. (1949). *Southern politics in state and nation*. New York: Knopf.

Kinsella, C., McTague, C., & Raleigh, K. N. (2015). Unmasking geographic polarization and clustering: A micro-scalar analysis of partisan voting behavior. *Applied Geography, 62*, 404–19.

Klinkner, P. A. (2004). Red and blue scare: The continuing diversity of the American electoral landscape. *The Forum, 2*(2), Article 2, 1–10.

Levendusky, M., & Pope, J. C. (2011). Red states vs. blue states: Going beyond the mean. *Public Opinion Quarterly, 75*(2), 227–48.

Martin, G. J., & Webster, S. (2017). Does residential sorting explain geographic polarization? Unpublished manuscript. http://polisci.emory.edu/faculty/gjmart2/papers/partisan_sorting_density.pdf

Massey, D. S., Rothwell, J., & Domina, T. (2009). The changing bases of segregation in the United States. *Annals of the American Academy of Political and Social Science, 626*(1), 74–90.

McGhee, E., & Krimm, D. (2009). Party registration and the geography of party polarization. *Polity, 41*(3), 345–67.

Motyl, M., Iyer, R., Oishi, S., Trawalter, S., & Nosek, B. A. (2014). How ideological migration geographically segregates groups. *Journal of Experimental Social Psychology, 51*, 1–14.

Mummolo, J., & Nall, C. (2017). Why partisans do not sort: The constraints on political segregation. *Journal of Politics, 79*(1), 45–59.

Myers, A. S. (2013). Secular geographical polarization in the American south: The case of Texas, 1996–2010. *Electoral Studies, 32*(1), 48–62.
Peck, J. (2005). Struggling with the creative class. *International Journal of Urban and Regional Research, 29*(4), 740–770.
Political Polarization in the American Public. (2014). Pew Research Center. http://www.people-press.org/files/2014/06/6-12-2014-Political-Polarization-Release.pdf
Rodden, J. (2010). The geographic distribution of political preferences. *Annual Review of Political Science, 13*(1), 321–40.
Schelling, T. C. (1971). Dynamic models of segregation. *Journal of Mathematical Sociology, 1*, 143–86.
Scher, R. K. (1997). *Politics in the New South: Republicanism, race and leadership in the twentieth century* (2nd ed.). Armonk/New York: M. E. Sharpe.
Strickler, R. (2016). A "sorted" America? Geographic polarization and value overlap in the American electorate. *Social Science Quarterly, 97*(2), 439–457.
Sussell, J. (2013). New support for the big sort hypothesis: An assessment of partisan geographic sorting in California, 1992–2010. *PS: Political Science & Politics, 46*(04), 768–73.
Tesler, M. (2016). *Post-racial or most-racial? Race and politics in the Obama era.* Chicago: University of Chicago Press.

Chapter 4
Modeling the Political Geography of Presidential Voting

As Chap. 3 demonstrated, the modest increase in geographic polarization since 1976 is not historically unique. Americans are no more likely to live in landslide counties today than they were in previous eras. But the political geography of blue counties and red counties identified in Chap. 2 still requires an explanation. Why have these patterns persisted and only changed gradually over time? It is to this question, and to the application of a modeling approach for spatial dependence that we turn in this chapter.

One of the key implications of the analyses in Chap. 2 is that there has been a strong spatial structuring of county-level voting in American presidential elections. Support for the parties has not been distributed randomly with regard to spatial location. Instead, neighboring counties have exhibited spatially autocorrelated levels of partisan support with each other in each election.

Broadly speaking this spatial dependence may be produced by either of two sources (see also Darmofal 2015). On the one hand, it may be produced by direct behavioral diffusion between neighboring counties. In this case, counties are influenced directly by the partisan voting behavior of their neighboring counties. So, for example, strong Democratic voting in county i leads directly to strong Democratic voting in neighboring county j and vice versa. This could happen, for example, due to shared social networks across neighboring counties. In the spatial diffusion case, a true spatial process is operating and we need to model this behavioral diffusion or risk biased parameter estimates due to omitted variable bias.

Alternatively, neighboring counties may be completely atomistic entities, divorced from their neighbors and unaware of the partisan voting preferences in those neighboring counties. And yet they may still exhibit spatial autocorrelation with these neighboring counties' partisan voting. They may do so due to shared attributes that predict partisan voting. For example, atomistic neighboring counties with similar racial demographics may exhibit similar partisan voting patterns because racial demographics predict voting behavior. If the sources of partisan

D. Darmofal, R. Strickler, *Demography, Politics, and Partisan Polarization in the United States, 1828–2016*, Spatial Demography Book Series 2,
https://doi.org/10.1007/978-3-030-04001-7_4

voting are unknown or unmeasured, we may need to employ a spatial modeling approach to avoid biased standard errors and type I errors in inference.

The two principal sources of spatial dependence thus each lead to a distinct type of spatial model. On the one hand, behavioral diffusion is consistent with a spatial lag model:

$$y = \rho W y + X\beta + \varepsilon, \tag{4.1}$$

where y is an N by 1 vector of observations on the dependent variable, Wy is a spatially lagged dependent variable with spatial weights matrix W, X is an N by K matrix of observations on covariates, β is a K by 1 vector of coefficients on the covariates, ρ is the spatial autoregressive parameter for the spatially lagged dependent variable, and ε is an N by 1 vector of error terms. A positive and statistically significant ρ indicates positive spatial autocorrelation consistent with behavioral diffusion. A negative and statistically significant ρ indicates negative spatial autocorrelation in which units' behavior leads to diverging behavior for their neighbors. Finally, an insignificant ρ indicates that behavioral diffusion is not present.

Alternatively, if neighboring counties exhibit similar partisan voting due to shared attributes rather than behavioral interactions, we can model this spatial dependence in one of two ways. First, if the attributes can be included in the model as covariates, we can employ a standard modeling approach such as ordinary least squares (OLS) regression. Alternatively, if the attributes are unknown or we lack measures for them, we will need to employ an alternative, spatial error modeling approach, as the correlated, unmodeled sources of spatial dependence will then go into the error term.

A spatial error model takes the following form:

$$y = X\beta + \varepsilon$$

$$\varepsilon = \lambda W \varepsilon + \xi, \tag{4.2}$$

where ε is an N by 1 vector of error terms, $W\varepsilon$ is a spatially lagged error term with spatial weights matrix W, λ is the spatial autoregressive coefficient for the spatially lagged error term, and ξ is, typically, an independent and identically distributed error term. A positive and statistically significant λ term indicates spatial error dependence for neighboring units. A negative and statistically significant λ indicates negative spatial error dependence for neighboring units. Finally, an insignificant λ indicates that spatial error dependence is absent.

We will often have strong theoretical reasons for expecting that spatial dependence of the type diagnosed in Chap. 2 is produced by either a spatial lag process or a spatial error process. These theoretical expectations, however, should be checked with the use of diagnostics. Happily, diagnostics have been developed that point the researcher to whether a spatial lag or spatial error model should be estimated.

4.1 Diagnostics for Spatial Dependence

The diagnostic approach employed in this chapter focuses on a set of Lagrange multiplier diagnostics for spatial dependence. The first diagnostic we examine is an LM diagnostic for spatial lag dependence:

$$LM_{Lag} = [Ne'Wy/e'e]^2[N(WX\hat{\beta})'M(WX\hat{\beta})/e'e + \mathrm{tr}(W'W + W^2)]^{-1}, \quad (4.3)$$

with N denoting the number of observations, e the OLS residuals, $M = I - X(X'X)^{-1}X'$, $\hat{\beta}$ as the OLS estimate of β, tr denoting the matrix trace operator, and W the spatial weights matrix for the spatially lagged dependent variable (Anselin and Rey 1991, 119). This diagnostic is distributed as a χ^2 statistic with 1 degree of freedom.

Just as there is an LM diagnostic for spatial lag dependence, so also is there an LM diagnostic for spatial error dependence. This diagnostic takes the form:

$$LM_{Error} = [Ne'We/e'e]^2[\mathrm{tr}(W'W + W^2)]^{-1}, \quad (4.4)$$

where the notation is as in (4.4) with W now the spatial weights matrix for the spatially lagged errors (Anselin 1988). Both of these LM diagnostics, however, carry with them a significant problem. Both can diagnose spatial dependence when only the alternative form of spatial dependence is present.

In other words, the LM diagnostic for spatial lag dependence may be statistically significant even if only spatial error dependence is present. Likewise, the LM diagnostic for spatial error dependence may be statistically significant even if only spatial lag dependence is present. Happily, robust Lagrange multiplier diagnostics have been developed that are robust to the alternative form of spatial dependence.

When the alternative form of spatial dependence is present, the simple unidirectional LM tests converge to a noncentral χ^2 distribution featuring an additional noncentrality parameter. Recognizing this, Bera and Yoon (1993) developed modified Lagrange multiplier tests that account for the noncentrality parameter, and are, as a consequence, robust to misspecification of the form of spatial dependence. Anselin, Bera, Florax, and Yoon 1996 subsequently extended Bera and Yoon's modified LM tests to the diagnosis of spatial lag and spatial error dependence in OLS specifications. In essence, the spatial lag dependence is estimated in the diagnostic for lag dependence by accounting for spatial error dependence that may exist. Likewise, spatial error dependence is estimated in the diagnostic for error dependence by accounting for spatial lag dependence that may exist.[1]

The robust Lagrange multiplier diagnostic for spatial lag dependence developed by Anselin, Bera, Florax, and Yoon (1996, 83) takes the form:

[1]This discussion of the robust Lagrange multiplier diagnostics originally appeared in Darmofal (2015). *Spatial Analysis for the Social Sciences*. New York: Cambridge University Press. Reprinted with permission.

$$Robust\ LM_{Lag} = \frac{(e'Wy/s^2 - e'We/s^2)^2}{N\tilde{J}_{\rho\cdot\beta} - t}, \tag{4.5}$$

where $s^2 = \frac{e'e}{N}$, and $N\tilde{J}_{\rho\cdot\beta} = [t + (WX\beta)'M(WX\beta)/s^2]$, with $M = I - X(X'X)^{-1}X'$, and $t = \text{tr}(W'W + W^2)$. Employing the same notation, the robust LM diagnostic for spatial error dependence takes the form:

$$Robust\ LM_{Error} = [e'We/s^2 - t(N\tilde{J}_{\rho\cdot\beta})^{-1}(e'Wy/s^2)]^2/[t - t^2(N\tilde{J}_{\rho\cdot\beta})^{-1}]. \tag{4.6}$$

The problem that the non-robust LM diagnostics pose for modeling spatial dependence occurs when both diagnose spatial dependence. If only one of the diagnostics indicates spatial autocorrelation, the researcher should estimate the model indicated by this diagnostic. Thus, for example, if the non-robust Lagrange multiplier diagnostic for spatial lag dependence is significant and the non-robust Lagrange multiplier diagnostic for spatial error dependence is not, the author should estimate a spatial lag model (Anselin 2005). This is because the danger with the non-robust LM diagnostic is that they may pick up the alternative form of dependence when only it is present. If the non-robust LM diagnostic for spatial error dependence is not significant, we can be confident that spatial error dependence is absent.

Anselin's (2005) decision rule for spatial diagnostics argues that if both of the non-robust LM diagnostics are statistically significant, one should next consult the robust LM diagnostics. If one is statistically significant and the other is not, the researcher should estimate the model indicated by the significant robust LM diagnostic. Thus, for example, if the robust LM error diagnostic is significant and the robust LM lag diagnostic is not, the researcher should estimate a spatial error model. If both are statistically significant, one should estimate the model indicated by the larger statistic. We follow this decision rule in estimating the spatial models of county-level voting in this chapter.

4.2 Analysis of Voting Behavior, 1828–2016

Our analysis of influences on presidential voting behavior since the advent of Jacksonian democracy in 1828 rests on a common set of covariates for each of these 48 presidential elections. By employing a common set of independent variables in each election, we're able to identify how factors influencing presidential voting have changed over time. Because of the paucity of measures that go back nearly 200 years, this necessarily limits the covariates we can include in our models. However, we believe that the benefits of utilizing the same model for each of the 48 elections to examine how these factors have changed over time outweighs the necessary limitations in the types of variables we can utilize. Additionally, our models include covariates capturing some of the principal fault lines in American political history: race, immigration, and the rural-urban divide.

The dependent variable for our analysis of county-level voting in the United States is the same variable we mapped in Chap. 2, *Margin*, the county-level margin of victory. This variable measures the Democratic proportion of the vote minus the National Republican/Whig/Republican proportion of the vote at the county level. We model the county-level margin of victory using a set of demographic covariates.

Race has played a central role in America's political history (see, e.g., King and Smith 2005) and we model it with a county-level covariate. The variable *Proportion African American* measures the proportion of the county's population that was African American.[2] Were we interested in drawing individual-level inferences from this covariate, the ecological fallacy would preclude this (Robinson 1950). However, we are not, and are primarily interested in how the aggregate racial context of a county shapes its voting behavior. Because our interest is in how aggregate context shapes behavior, and because populations of in-groups and out-groups in a county can produce different aggregate relationships, we do not have directional expectations for this variable.

In addition to race, immigration has also played a critical role in American political development (see, e.g., Mink 1990). We capture the effects of local immigrant populations with the variable, *Proportion Foreign-Born*, which measures the proportion of the county's population that was foreign-born. For the same reasons as with the aggregate-level racial variable, we do not have a directional expectation for the *Proportion Foreign-Born* covariate.

We are also interested in how population density affected partisan voting. To capture this, we included the covariate *Population Density*, which measures the population density of the county.[3] This measure captures differences in voting between more urban and more rural locales. In doing so, it captures how differing interests and differing social networks in more and less densely populated counties affected voting behavior.

Rather than employing a panel data analysis with our unbalanced data (in which counties enter and, during the Civil War and Reconstruction, exit) the Electoral College over time, we instead employ separate year by year analyses. This also has the advantage of providing more clarity as to the effects of our covariates on partisan voting in each election examined in Chap. 2, instead of imposing a pooling assumption and estimating average effects of our covariates over time.

[2]Including a variable for the proportion white in the models would be superfluous as it is negatively correlated with *Proportion African American* at a value of −0.96 for the full period of our analysis and at a level of −.99 or higher in most elections until the 1920s. In short, whites and African Americans accounted for nearly all of the population in most counties well into the twentieth century.

[3]This variable is interpolated for counties in early elections.

4.3 The Second American Party System

Our analysis of the correlates of county-level voting in presidential elections begins with an analysis of the Second American Party System. The results for the elections in this period are presented in Table 4.1. We note, first, that in all elections in this party system, and, indeed, in all of the presidential elections from 1828 to the present, the spatial diagnostics indicated that a spatial model was preferred over OLS estimation of a standard model with no spatial dependence. The maps in Chap. 2 had demonstrated the strong spatial structuring of partisan voting and this dependence persists in the presence of our covariates.

We employed Anselin's decision rule and Lagrange multiplier diagnostics to determine whether a spatial lag model or a spatial error model was appropriate. Elections in which a spatial lag model was the preferred model specification are indicated by an estimate for ρ, the spatial lag parameter. Elections in which a spatial error model was the preferred specification are indicated by an estimate for λ, the spatial autoregressive error parameter. As can be seen from Table 4.1, the spatial lag model was the preferred model specification in 5 of the 7 presidential elections during the Second Party System, with the spatial error model preferred in the other two elections (the elections of 1840 and 1844).

The spatial lag parameter, ρ, is strongly significant and positive in each presidential election in the Second Party System, reaching statistical significance at a $p < .001$ level. The values for ρ are also unusually high, never below .75 in any election in which the spatial lag model was estimated. There is a strong positive spatial lag process in the data in most elections in this party system, consistent with

Table 4.1 Spatial models of county-level voting in the second party system, 1828–1852

Year	ρ	λ	Constant	Proportion African American	Proportion Foreign Born	Population Density
1828	0.86***		0.03***	0.04	−0.25	0.02
	(0.02)		(0.01)	(0.03)	(0.13)	(0.02)
1832	0.86***		0.04***	0.06*	−0.36	0.02
	(0.02)		(0.01)	(0.03)	(0.19)	(0.15)
1836	0.75***		0.02	−0.02	0.03	−0.12
	(0.02)		(0.01)	(0.04)	(0.23)	(0.16)
1840		0.77***	0.02	−0.23***	−0.31	0.02
		(0.02)	(0.03)	(0.06)	(0.28)	(0.14)
1844		0.74***	0.13***	−0.23***	−0.53**	0.02
		(0.02)	(0.03)	(0.05)	(0.21)	(0.11)
1848	0.80***		0.04***	−0.06*	−0.18	−0.07
	(0.02)		(0.01)	(0.03)	(0.10)	(0.07)
1852	0.78***		0.05***	−0.00	−0.13*	0.02
	(0.02)		(0.01)	(0.02)	(0.06)	(0.06)

$^{*}p < 0.05$, $^{**}p < 0.01$, $^{***}p < 0.001$

a positive spatial diffusion process in which partisan voting in a county directly shapes partisan voting in neighboring counties (and vice versa).

In the two elections for which a spatial error model was preferred, the λ parameter is also strongly significant and positive. This means that there is spatial dependence in the error terms. In other words, absent a diffusion process, the covariates in the model are not capturing other omitted sources of spatial dependence.

We can see that the *Proportion African American* covariate is statistically significant in the majority of presidential elections in the Second Party System. In 1832, it is significant and positively signed. Counties with larger African American populations in 1832 had larger margins for Jackson over Clay. This is, of course, a county-level covariate and this doesn't mean that African Americans were voting for Jackson over Clay during a time period in which many African Americans couldn't vote. Indeed, as we will see later in this chapter, this covariate becomes consistently positive and significant during the Jim Crow era of the solid Democratic South, in which counties with large populations of African Americans were voting Democratic because of the disenfranchisement of these would-be likely Republican voters.

After an election in which the proportion of African Americans was unrelated to partisan voting, this covariate once again predicted voting behavior in the 1840, 1844, and 1848 presidential elections. Now, however, it was negatively related to the *Margin* variable. Counties with larger African American populations were more pro-Whig than counties with smaller African American populations.

Of course we know that the Second Party System collapsed because it was unable to address the central issues of race and slavery that were of increased interest to voters. It is instructive, therefore, that this race covariate was a significant predictor of county-level voting in a majority of elections in this party system. Race was influencing voting behavior even when political elites were unwilling to effectively address the issue.

Race was a more consequential influence on voting behavior during the Second Party System than was the presence of immigrants. In only two elections during the second wave of immigration to the United States did the *Proportion Foreign-Born* covariate reach statistical significance. These were the elections of 1844 and 1852. In both elections, larger foreign-born populations were associated with more pro-Whig voting.

Finally, the Second Party System was not marked by a strong rural-urban partisan cleavage. Population density did not predict county-level voting in any election in the Second System. This might come as a surprise to modern observers, who have seen such a strong rural-urban divide in recent elections. But in fact, such a partisan divide didn't become commonplace in our politics until the 1920s, as we will discuss later in this chapter.

4.4 The Third American Party System

The issues of race and slavery that went unaddressed in the Second Party System, despite shaping voting behavior during that party system, were arguably the central political cleavage of the Third American Party System. In order to gauge the effects of racial demographics and the other covariates during the Third Party System, we estimated the same model specifications for each election during this party system. The results for the 1856–1880 presidential elections are presented in Table 4.2 while the results for the 1884–1892 presidential elections are presented in Table 4.3.

We turn first, again, to the spatial terms. As in the Second Party System, the spatial dependence in the Third Party System was largely reflective of a spatial diffusion process. Of the 10 presidential elections during this system, eight were best modeled via spatial lag models. And in each of these eight elections, the spatial autoregressive parameter, ρ, was highly significant (all at a $p < .001$ level) and positively signed. Again the ρ estimates have large positive values, indicating a strong positive spatial lag process in the data.

The two elections for which spatial error models were appropriate were the 1864 and 1872 elections. In both elections, the spatial error parameter, λ, is significant at a $p < .001$ level and is positively signed. Omitted covariates produced spatial dependence in the errors in these elections. It is interesting to see that the 1864 election that took place during the Civil War was one of the few elections in the nineteenth century for which a spatial error model was the preferred model. It is possible that the war cut through the usual lines of communication across locales that promoted spatial diffusion in voting behavior in other elections. Likewise, the Reconstruction Era election of 1872 may have also been one in which the diffusion process was interrupted due to the unusual circumstances of that election.

Table 4.2 Spatial models of county-level voting in the third party system, 1856–1880

Year	ρ	λ	Constant	Proportion African American	Proportion Foreign Born	Population Density
1856	0.87***		0.05***	0.07***	−0.17***	0.04
	(0.01)		(0.01)	(0.02)	(0.04)	(0.04)
1860	0.83***		0.00	0.04*	−0.11**	0.05
	(0.01)		(0.01)	(0.02)	(0.04)	(0.03)
1864		0.77***	−0.22***	1.18***	0.21	0.05
		(0.02)	(0.04)	(0.18)	(0.15)	(0.06)
1868	0.77***		−0.00	0.09*	−0.18**	0.09
	(0.02)		(0.01)	(0.04)	(0.06)	(0.05)
1872		0.82***	0.12***	−0.58***	−0.80***	0.06
		(0.01)	(0.03)	(0.05)	(0.15)	(0.05)
1876	0.80***		0.11***	−0.16***	−0.54***	0.06
	(0.01)		(0.01)	(0.03)	(0.06)	(0.03)
1880	0.81***		0.09***	−0.13***	−0.49***	0.05
	(0.01)		(0.01)	(0.02)	(0.06)	(0.03)

*$p < 0.05$; **$p < 0.01$; ***$p < 0.001$

Table 4.3 Spatial models of county-level voting in the third party system, 1884–1892

Year	ρ	λ	Constant	Proportion African American	Proportion Foreign Born	Population Density
1884	0.79***		0.08***	−0.11***	−0.44***	0.06
	(0.01)		(0.01)	(0.02)	(0.05)	(0.03)
1888	0.81***		0.05***	−0.02	−0.31***	0.07
	(0.01)		(0.01)	(0.02)	(0.05)	(0.05)
1892	0.83***		0.02***	0.09***	−0.22***	0.11*
	(0.01)		(0.01)	(0.02)	(0.04)	(0.05)

*$p < 0.05$; **$p < 0.01$; ***$p < 0.001$

The covariate, *Proportion African American*, is significant in each election during the Third Party System. Race clearly was a central issue in the battles between the Republicans and the Democrats during this system. The signs of the coefficients suggest a very interesting story. In the pre-Civil War elections, the war elections, and the immediate post-War election, the sign of the effect is positive. Larger proportions of African Americans in a county were associated with stronger support for the Democrats.

This isn't a surprising finding. With black enfranchisement non-existent where slavery (and, increasingly, Democrats) were entrenched, the counties with large black populations tended to oppose the new anti-slavery Republican Party in the pre-War elections. This continued in the Civil War election of 1864 and in the first Reconstruction Era election of 1868. By 1872, however, as Reconstruction proceeded, the sign flipped and became negative. Counties with larger African American populations showed more Republican support. We cannot, of course, know which voters at the individual level were producing the aggregate pro-Republican effect. However, the result is consistent with what we might expect to see during a period in which African American males finally received the right to vote nationally through the 14th and 15th Amendments to the Constitution.

The relationship between the proportion African American in a county and support for Republicans continued after the end of Reconstruction in 1876. In that election as well as the two succeeding elections, the pattern held. This ended, however, in 1888, when there was no relationship between this covariate and the margin of the vote in the county. And a new pattern would emerge beginning in 1892, at the end of the Third Party System, as Jim Crow began to set in, with counties with larger African American populations having larger Democratic margins of victory. This relationship would be one of the defining features of the Fourth Party System in which the Jim Crow regime played such a central role.

The presence of immigrants in a county played a much larger role in shaping partisan voting during the Third Party System than it had during the Second Party System. During the Second Party System, the median percentage foreign-born in counties had been one percent of the population. During the Third Party System, it rose to six percent. And the size of the immigrant population in a county influenced partisan voting in all but one presidential election during the Third Party System.

In all but one election during this period, *Proportion Foreign-Born* was negatively related to Democratic support in the county. Counties with larger immigrant populations were more likely to vote for the more nativist Republican Party.

Finally, urbanity and rurality were again unrelated to partisan voting during most elections in the Third Party System. Only in the 1892 election was it statistically significant, with less densely populated counties having larger Democratic margins. The rural-urban divide that is so central to the partisan divide in contemporary America simply didn't shape partisan voting during the nineteenth century.

4.5 The Fourth American Party System

Spatial dependence was once again ubiquitous in the presidential elections of the Fourth American Party System. In each election from 1896 through 1928, either a spatial lag or a spatial error model was preferred over a non-spatial model (Table 4.4). And once again, positive spatial diffusion in voting behavior dominated during this period. In seven of the nine elections during this period, a spatial lag model was preferred over a spatial error model.

The spatial diffusion was again significant at a $p < .001$ level and took extremely large values. The autoregressive parameter for the spatial diffusion process, ρ, was never below .79 during the Fourth Party System. Likewise, in the two elections for which a spatial error model was preferred (1920 and 1928), the spatial error dependence was at very high levels of .88 and .77, respectively. Partisan voting, in short, was strongly spatially structured in the Fourth American Party System even with covariates included in the models.

Table 4.4 Spatial models of county-level voting in the fourth party system, 1896–1928

Year	ρ	λ	Constant	Proportion African American	Proportion Foreign Born	Population Density
1896	0.85***		0.04***	0.02	−0.17***	0.02
	(0.01)		(0.01)	(0.02)	(0.04)	(0.05)
1900	0.79***		0.02*	0.16***	−0.25***	0.06
	(0.01)		(0.01)	(0.02)	(0.05)	(0.04)
1904	0.81***		0.00	0.23***	−0.35***	0.08*
	(0.01)		(0.01)	(0.02)	(0.05)	(0.03)
1908	0.82***		0.01	0.18***	−0.22***	0.05
	(0.01)		(0.01)	(0.02)	(0.04)	(0.03)
1912	0.82***		0.04***	0.18***	−0.07*	0.02
	(0.01)		(0.01)	(0.02)	(0.03)	(0.02)
1916	0.82***		0.01*	0.23***	−0.06	0.02
	(0.01)		(0.01)	(0.02)	(0.04)	(0.02)
1920		0.88***	−0.10***	0.63***	−0.65***	0.01
		(0.01)	(0.03)	(0.05)	(0.14)	(0.02)

$^{*}p < 0.05$; $^{**}p < 0.01$; $^{***}p < 0.001$

Table 4.5 Spatial models of county-level voting in the fourth party system, 1924–1928

Year	ρ	λ	Constant	Proportion African American	Proportion Foreign Born	Population Density
1924	0.83***		0.00	0.25***	−0.46***	0.04**
	(0.01)		(0.01)	(0.02)	(0.06)	(0.01)
1928		0.77***	−0.18***	0.80***	−0.33*	0.05*
		(0.01)	(0.02)	(0.05)	(0.17)	(0.02)

$^{*}p < 0.05$; $^{**}p < 0.01$; $^{***}p < 0.001$

Race continued to significantly structure voting behavior during the Fourth Party System. In every election except 1896, *Proportion African American* was a statistically significant predictor of county-level voting behavior. And in each election in which it was significant, larger African American populations were associated with stronger support for the Democratic Party. Of course, this was an era in which Jim Crow politics dominated in the Solid Democratic South and thus we shouldn't be surprised that in counties in which conservative Democrats disenfranchised large populations of African Americans, support for the Democratic Party resulted among those whites still able to vote.

The Fourth Party System occurred during a period of significant immigration to the United States. It is not surprising, therefore, that immigrant populations continued to influence aggregate voting in presidential elections. In eight of the nine presidential elections during the Fourth Party System, *Proportion Foreign-Born* had a negative effect on the margin of victory variable. Once again, the larger the local immigrant population, the more the county voted for the Republican Party.

Once again, the rural-urban divide generally did not predict counties' partisan voting during the Fourth Party System. 1904 was the lone election from 1896 until the 1920s in which population density did predict counties' voting behavior. Near the end of this party system, however, population density did begin to be related to voting. In both 1924 and 1928, the population density variable was related to the margin of victory in the county (Table 4.5). In both elections, more densely populated counties were more likely to support the Democratic Party. This would foreshadow a trend that would continue for several decades (Table 4.6).

4.6 The Fifth American Party System

Spatial diffusion in voting behavior continued to dominate in the Fifth American Party System. In all but two elections in the New Deal party system, there is evidence consistent with spatial diffusion. And again, the autoregressive diffusion parameter, ρ, takes on very high positive values in each of these elections and is again significant at a $p < .001$ level. In all elections in which a spatial lag model was preferred, ρ took on values of .82 or larger. Only in 1936 and 1960 was a spatial error model preferred. In both elections, the λ parameter was positively signed and significant at a $p < .001$ level. In both elections, λ was .80 or higher (Table 4.7).

Table 4.6 Spatial models of county-level voting in the fifth party system, 1932–1956

Year	ρ	λ	Constant	Proportion African American	Proportion Foreign Born	Population Density
1932	0.86***		0.05***	0.17***	−0.22***	0.05***
	(0.01)		(0.01)	(0.02)	(0.06)	(0.01)
1936		0.88***	0.25***	0.51***	0.23	0.04*
		(0.01)	(0.02)	(0.04)	(0.18)	(0.02)
1940	0.84***		0.01	0.27***	−0.17*	0.06***
	(0.01)		(0.01)	(0.02)	(0.08)	(0.02)
1944	0.84***		−0.01	0.29***	−0.08	0.06***
	(0.01)		(0.01)	(0.02)	(0.09)	(0.02)
1948	0.85***		0.02**	0.08***	−0.28**	0.05***
	(0.01)		(0.01)	(0.02)	(0.10)	(0.02)
1952	0.82***		−0.03***	0.18***	−0.40***	0.06***
	(0.01)		(0.01)	(0.02)	(0.11)	(0.01)
1956	0.82***		−0.03***	0.19***	−0.40***	0.03**
	(0.01)		(0.01)	(0.02)	(0.11)	(0.01)

*$p < 0.05$; **$p < 0.01$; ***$p < 0.001$

Table 4.7 Spatial models of county-level voting in the fifth party system, 1960–1964

Year	ρ	λ	Constant	Proportion African American	Proportion Foreign Born	Population Density
1960		0.80***	−0.10***	0.50***	0.21	0.02
		(0.01)	(0.02)	(0.04)	(0.23)	(0.01)
1964	0.86***		0.01*	−0.06***	0.42***	0.02*
	(0.01)		(0.01)	(0.02)	(0.10)	(0.01)

*$p < 0.05$; **$p < 0.01$; ***$p < 0.001$

Race continued to be an important predictor of county-level voting in the New Deal party system. In each of the first eight elections in this party system, *Proportion African American* was statistically significant and positively signed. In each election in the era, counties with larger African American populations had larger Democratic margins of victory. Again, with county-level rather than individual-level data it is impossible to determine which voters were producing these aggregate results. In fact, it's possible that multiple different individual-level processes were producing the aggregate results. In the still Solid Democratic South, conservative Democratic elites continued to disenfranchise African Americans and the whites who were still able to vote in these counties were, we know from the historical record, voting Democratic. In other locales, particularly in the North, the Democrats likely started picking up some African American votes, given the appeal of the New Deal policy agenda and the specific outreach to African American voters by First Lady Eleanore Roosevelt. In each of these elections, the positive relationship between *Proportion African American* and *Margin* was highly significant, at a $p < .001$ level.

Table 4.8 Spatial models of county-level voting in the sixth party system, 1968–1992

Year	ρ	λ	Constant	Proportion African American	Proportion Foreign Born	Population Density
1968		0.78***	−0.19***	0.82***	0.92***	0.02
		(0.01)	(0.01)	(0.03)	(0.17)	(0.01)
1972	0.77***		−0.11***	0.11***	0.56***	0.03**
	(0.01)		(0.01)	(0.02)	(0.09)	(0.01)
1976		0.81***	0.01	0.60***	−0.51**	0.05**
		(0.01)	(0.01)	(0.04)	(0.17)	(0.02)
1980		0.83***	−0.20***	0.92***	−0.13	0.07***
		(0.01)	(0.01)	(0.03)	(0.15)	(0.02)
1984		0.83***	−0.20***	0.89***	−0.09	0.07***
		(0.01)	(0.01)	(0.03)	(0.15)	(0.02)
1988		0.84***	−0.20***	1.03***	−0.00	0.09***
		(0.01)	(0.01)	(0.04)	(0.15)	(0.02)
1992		0.82***	−0.06***	0.85***	0.05	0.08***
		(0.01)	(0.01)	(0.03)	(0.13)	(0.02)

$^{*}p < 0.05$; $^{**}p < 0.01$; $^{***}p < 0.001$

An interesting change, however, occurred in the 1964 election. In this election, the Democratic candidate and incumbent president, Lyndon Johnson, took a strong civil rights position in support of the Civil Rights Act of 1964. His Republican opponent, Barry Goldwater, reversed the historical Republican support for civil rights legislation by opposing the Civil Rights Act on states' rights grounds. As a consequence, Goldwater won five states in the Deep South (Alabama, Georgia, Louisiana, Mississippi, and South Carolina) with large African American populations, in which these populations were still largely disenfranchised (the Voting Rights Act of 1965 would not be enacted until a year later). It is not surprising, therefore, that the sign on the *Proportion African American* covariate flipped in 1964, with a strong negative relationship between it and the Democratic margin of victory in the county. This effect was again significant at a $p < .001$ level.

Proportion Foreign-Born continued to be related to the partisan vote in the Fifth Party System. In six of the nine elections in the system it had a significant effect on the margin variable. In the first five of these elections, the relationship was a negative one. Counties with larger foreign-born populations exhibited less support for the Democratic nominees. This sign also flipped in 1964, to a positive effect, with counties with larger immigrant populations exhibiting more support for Lyndon Johnson.

In the Fifth American Party System, for the first time in the era of mass voter participation in the U.S., the rural-urban divide consistently shaped voting behavior. In all but one election in the period, *Population Density* had a significant relationship to the margin variable. And in each of these elections, the relationship was a positive one. Consistent with the current relationship, more densely populated counties exhibited stronger support for the Democratic candidates.

One of the key take-home points from this analysis is that examining the correlates of voting on an election-by-election basis can be illuminating for identifying developments in American electoral history. A pooled time-series cross-sectional analysis, even a spatial one, would miss the over-time changes in the effects of covariates such as *Proportion Foreign Born* and *Population Density*. By examining the factors that influenced each election, we can identify changes in covariates' effects over-time as well as anomalous elections.

4.7 The Sixth American Party System

The first important feature regarding the Sixth American Party System is the spatial process that has dominated during this party system. Unlike the Second through Fifth Party Systems, in which a spatial diffusion process was the dominant spatial process, spatial error processes have been the operative process in all but one presidential election during the Sixth Party System (see Tables 4.8 and 4.9). The 1972 Nixon landslide is the only presidential election during this system in which a spatial diffusion process was in effect. In all the other elections in the system, spatial error dependence predominated. Although it is impossible to determine why spatial diffusion no longer produces spatial dependence in partisan voting behavior from this model specification, it is instructive that the Sixth American Party System has taken place during a period of fraying in Americans' social connections to each other (Putnam 2000). With the decline of social attachments to other citizens, it is perhaps not surprising that partisan preferences do not as easily diffuse geographically as they once did.

Just as in the Fifth American Party System, a larger African American population in a county has been associated with stronger support for the Democratic Party in the current party system. In fact, this positive relationship has existed in each presidential election in the Sixth American Party System. Larger immigrant populations were associated with stronger Democratic support in the 1968 and 1972 elections, but larger Republican support in the 1976 election. After this, there was no relationship between the proportion foreign-born in a county and its partisan voting for four presidential elections. This changed with Clinton's re-election campaign in 1996 and has remained positive and significant in each election since (Table 4.9). Larger immigrant populations in counties are now consistently associated with stronger Democratic support.

The urban-rural divide that began in the late Fourth Party System and continued in the Fifth Party System has continued also in the current Sixth Party System. In fact in each election in this party system except the first (in 1968), more densely populated counties have supported the Democratic Party more strongly. This relationship is now in its tenth consecutive decade and remains one of the defining features of our national politics.

Table 4.9 Spatial models of county-level voting in the sixth party system, 1996–2016

Year	ρ	λ	Constant	Proportion African American	Proportion Foreign Born	Population Density
1996		0.81***	−0.07***	0.82***	0.35*	0.08***
		(0.01)	(0.01)	(0.03)	(0.17)	(0.02)
2000		0.82***	−0.26***	0.96***	0.01***	0.10***
		(0.01)	(0.01)	(0.03)	(0.00)	(0.02)
2004		0.87***	−0.33***	1.34***	0.78***	0.09***
		(0.01)	(0.02)	(0.04)	(0.08)	(0.02)
2008		0.88***	−0.30***	1.53***	1.08***	0.09***
		(0.01)	(0.02)	(0.04)	(0.08)	(0.02)
2012		0.89***	−0.37***	1.67***	1.08***	0.09***
		(0.01)	(0.02)	(0.04)	(0.07)	(0.02)
2016		0.85***	−0.51***	1.69***	1.58***	0.12***
		(0.01)	(0.02)	(0.04)	(0.08)	(0.02)

$^{*}p < 0.05$; $^{**}p < 0.01$; $^{***}p < 0.001$

4.8 The Interaction of Racial and Immigrant Demographics

Thus far we have examined, among other questions, whether racial and immigrant demographics in counties shaped their voting behavior. Both racism and opposition to immigrants have been common themes throughout American history. But the electoral effects of race and immigration do not necessarily operate in isolation from each other. Instead, it may well be that there is an interactive effect with the joint presence of large populations of African Americans and immigrants in a county having an additional, multiplicative effect on voting behavior. It is to this question that we turn next.

To examine this, we again estimated models for each presidential election. The covariates in the models are the same as in the preceding analysis, with the exception of an additional *Proportion African American x Proportion Foreign-Born* interaction term. Our dependent variable is again the margin of victory variable. As with the previous models, we estimates these models for each election and report them by party system.

The results for the models with the *Proportion African American x Proportion Foreign-Born* interaction term for the Second American Party System are reported in Table 4.10. As we can see, there were no consistent effects for this interaction during the Second Wave of Immigration to the United States. In only two elections (1832 and 1848) was the interaction statistically significant and the interaction took different signs in these years, with a negative effect in 1832 and a positive one in 1848.[4]

[4]As Brambor et al. (2006) recognize, an interaction can have a significant effect for a portion of its range even if the interaction term itself is statistically insignificant.

Table 4.10 Spatial models of county-level voting in the second party system, 1828–1852

Year	ρ	λ	Constant	Proportion African American	Proportion Foreign Born	Proportion African American × Proportion Foreign Born	Population Density
1828	0.86***		0.03***	0.04	−0.21	−2.01	0.02
	(0.02)		(0.01)	(0.03)	(0.13)	(2.46)	(0.02)
1832	0.85***		0.04***	0.10**	−0.04	−5.80**	0.03
	(0.02)		(0.01)	(0.03)	(0.22)	(2.10)	(0.15)
1836	0.75***		0.03*	−0.05	−0.25	3.58	−0.12
	(0.02)		(0.01)	(0.04)	(0.28)	(1.92)	(1.62)
1840		0.77***	0.02*	−0.23***	−0.29	−0.17	0.02
		(0.02)	(0.03)	(0.07)	(0.32)	(2.03)	(0.14)
1844		0.74***	0.13***	−0.23***	−0.55*	0.11	0.02
		(0.02)	(0.03)	(0.06)	(0.23)	(1.28)	(0.11)
1848	0.78***		0.05***	−0.13***	−0.44***	2.22***	−0.03
	(0.02)		(0.01)	(0.03)	(0.12)	(0.56)	(0.07)
1852	0.78***		0.05***	−0.01	−0.16*	0.60	0.02
	(0.02)		(0.01)	(0.03)	(0.07)	(0.49)	(0.06)

$^{*}p < 0.05$; $^{**}p < 0.01$; $^{***}p < 0.001$

The inclusion of the interaction term did not affect the conclusions drawn regarding population density or the spatial terms the model. Once again, population density is insignificant in all elections in the Second Party System. Where there was a positive and significant ρ parameter for the spatial diffusion process in the models without the interaction term, there was a positive and significant ρ in the models with the interaction term included. The same held true for the two elections in the Second Party System with significant positive spatial error dependence.

In the Third American Party System, the interaction effect of large African American and immigrant populations began to exert a more consistent effect on partisan voting (see Tables 4.11 and 4.12). In 1860, 1868, and 1872 the interaction term was statistically significant. In each of these elections, this interaction effect was positively signed. Of course, with aggregate data alone we cannot determine which citizens were producing this effect. This effect disappeared in subsequent elections and was even replaced with a negative interaction effect near the end of the Third Party System in 1888.

The inclusion of the interaction term did not affect the inferences drawn regarding the population density variable. Once again it was insignificant in all but the last election in the Third Party System where, once again, it was significant and positively signed. The diffusion and error processes likewise remained largely unchanged. The lone exception was the 1872 election, which was reflective of a positive spatial error process in the model without the interaction and which was reflective of a positive spatial lag process in the model with the interaction included.

As Tables 4.13 and 4.14 show, the interaction term was statistically significant in 5 of the 9 elections in the Fourth American Party System. In each of these five

Table 4.11 Spatial models of county-level voting in the third party system, 1856–1880

Year	ρ	λ	Constant	Proportion African American	Proportion Foreign Born	Proportion African American × Proportion Foreign Born	Population Density
1856	0.87***		0.05***	0.05*	−0.19***	0.74	0.04
	(0.01)		(0.01)	(0.02)	(0.04)	(0.39)	(0.04)
1860	0.83***		0.00	0.02	−0.13***	0.90*	0.05
	(0.01)		(0.01)	(0.02)	(0.04)	(0.39)	(0.03)
1864		0.77***	−0.23***	1.30***	0.25	−1.76	0.05
		(0.02)	(0.04)	(0.22)	(0.17)	(2.08)	(0.06)
1868	0.76***		−0.00	0.06	−0.21**	2.32*	0.09
	(0.02)		(0.01)	(0.04)	(0.07)	(1.05)	(0.05)
1872	0.79***		−0.07***	−0.21***	−0.53***	1.61*	0.06
	(0.01)		(0.01)	(0.03)	(0.06)	(0.738)	(0.038)
1876	0.80***		0.11***	−0.16***	−0.54***	−0.12	0.06
	(0.01)		(0.01)	(0.03)	(0.06)	(0.73)	(0.03)
1880	0.81***		0.09***	−0.13***	−0.50***	0.54	0.05
	(0.01)		(0.01)	(0.03)	(0.06)	(0.67)	(0.03)

$^{*}p < 0.05$; $^{**}p < 0.01$; $^{***}p < 0.001$

Table 4.12 Spatial models of county-level voting in the third party system, 1884–1892

Year	ρ	λ	Constant	Proportion African American	Proportion Foreign Born	Proportion African American × Proportion Foreign Born	Population Density
1884	0.79***		0.08***	−0.09***	−0.43***	−0.89	0.06
	(0.01)		(0.01)	(0.02)	(0.05)	(0.64)	(0.03)
1888	0.81***		0.05***	−0.00	−0.30***	−1.34*	0.08
	(0.01)		(0.01)	(0.02)	(0.05)	(0.61)	(0.05)
1892	0.82***		0.02**	0.08***	−0.23***	0.94	0.11*
	(0.01)		(0.01)	(0.02)	(0.04)	(0.60)	(0.05)

$^{*}p < 0.05$; $^{**}p < 0.01$; $^{***}p < 0.001$

elections, this interaction effect was negatively signed. Once again, the inclusion of the interaction term did not affect inferences on the population density covariate. Where it was significant (and positive) in the models without interactions it was also significant (and positive) in the models with them. Likewise, the inferences regarding spatial diffusion and spatial error processes remained unchanged in the two sets of models.

As Table 4.15 shows, the significant negative interaction effect continued in the first two elections of the Fifth American Party System. We can gain a sense of the substantive impact of this interaction effect by examining how changes in the sizes of local African American and immigrant populations affected the predicted margin in 1932. With both terms set at their means, the predicted margin of victory for Franklin Roosevelt was 29.6 percentage points. Shifting just *Proportion*

Table 4.13 Spatial models of county-level voting in the fourth party system, 1896–1920

Year	ρ	λ	Constant	Proportion African American	Proportion Foreign Born	Proportion African American × Proportion Foreign Born	Population Density
1896	0.85***		0.04***	0.04	−0.16***	−1.91**	0.03
	(0.01)		(0.01)	(0.02)	(0.04)	(0.69)	(0.05)
1900	0.79***		0.02*	0.17***	−0.24***	−1.52*	0.07
	(0.01)		(0.01)	(0.02)	(0.05)	(0.73)	(0.04)
1904	0.81***		0.00	0.23***	−0.35***	−0.47	0.08*
	(0.01)		(0.01)	(0.02)	(0.05)	(0.64)	(0.03)
1908	0.82***		0.01	0.18***	−0.23***	0.04	0.05
	(0.01)		(0.01)	(0.02)	(0.05)	(0.65)	(0.03)
1912	0.82***		0.04***	0.19***	−0.06***	−1.02*	0.02
	(0.01)		(0.01)	(0.02)	(0.03)	(0.52)	(0.02)
1916	0.83***		0.01*	0.24***	−0.05	−0.75	0.02
	(0.01)		(0.01)	(0.02)	(0.05)	(0.61)	(0.02)
1920		0.88***	−0.12***	0.69***	−0.46**	−4.23***	0.02
		(0.01)	(0.03)	(0.05)	(0.15)	(1.19)	(0.02)

$^{*}p < 0.05$; $^{**}p < 0.01$; $^{***}p < 0.001$

Table 4.14 Spatial models of county-level voting in the fourth party system, 1924–1928

Year	ρ	λ	Constant	Proportion African American	Proportion Foreign Born	Proportion African American × Proportion Foreign Born	Population Density
1924	0.83***		0.00	0.26***	−0.44***	−0.98	0.04**
	(0.01)		(0.01)	(0.02)	(0.06)	(0.80)	(0.01)
1928		0.77***	−0.19***	0.92***	−0.01	−11.66***	0.09***
		(0.01)	(0.02)	(0.05)	(0.18)	(1.98)	(0.02)

$^{*}p < 0.05$; $^{**}p < 0.01$; $^{***}p < 0.001$

African American to one standard deviation above its mean reduced the predicted FDR margin to 26.7 points. Shifting just the foreign-born variable to one standard deviation above its mean reduced the predicted Democratic margin to 24.2 points. Shifting both *Proportion Foreign-Born* and *Proportion African-American* to one standard deviation above their means reduced the predicted Democratic margin in 1932 to 11.7 points.

However, as both Tables 4.15 and 4.16 show, in most elections during this party system, there was no significant effect for this interaction term. In the Fifth Party System the *Population Density* covariate was significant and positive in all but the second-to-last election in the party system in the models without interaction terms. This was only slightly changed in the models with interactions, with the covariate also insignificant in the final election in the party system. There was also only one change on inferences regarding lag vs. error processes across the two sets of models. Where a positive lag process was operative in 1932 in the model without

Table 4.15 Spatial models of county-level voting in the fifth party system, 1932–1956

Year	ρ	λ	Constant	Proportion African American	Proportion Foreign Born	Proportion African American × Proportion Foreign Born	Population Density
1932		0.90***	0.32***	0.37***	−0.14	−10.67***	0.09***
		(0.01)	(0.02)	(0.04)	(0.17)	(1.66)	(0.02)
1936		0.89***	0.24***	0.55***	0.41*	−4.82**	0.07***
		(0.01)	(0.02)	(0.05)	(0.19)	(1.71)	(0.02)
1940	0.84***		0.01	0.27***	−0.18*	−0.61	0.05***
	(0.01)		(0.01)	(0.02)	(0.08)	(1.35)	(0.02)
1944	0.84***		−0.01	0.28***	−0.11	1.78	0.06***
	(0.01)		(0.01)	(0.03)	(0.09)	(1.40)	(0.02)
1948	0.85***		0.02**	0.06**	−0.35***	2.88*	0.04*
	(0.01)		(0.01)	(0.02)	(0.11)	(1.44)	(0.02)
1952	0.82***		−0.03***	0.17***	−0.48**	2.53	0.05**
	(0.01)		(0.01)	(0.02)	(0.12)	(1.45)	(0.02)
1956	0.82***		−0.03***	0.18***	−0.42***	1.21	0.03**
	(0.01)		(0.01)	(0.02)	(0.11)	(0.78)	(0.01)

$^{*}p < 0.05$; $^{**}p < 0.01$; $^{***}p < 0.001$

Table 4.16 Spatial models of county-level voting in the fifth party system, 1960–1964

Year	ρ	λ	Constant	Proportion African American	Proportion Foreign Born	Proportion African American × Proportion Foreign Born	Population Density
1960		0.80***	−0.09***	0.47***	−0.01	2.47	0.01
		(0.01)	(0.02)	(0.05)	(0.27)	(1.59)	(0.01)
1964	0.85***		0.20***	−0.13***	0.02	6.90***	0.00
	(0.01)		(0.01)	(0.02)	(0.12)	(1.08)	(0.01)

$^{*}p < 0.05$; $^{**}p < 0.01$; $^{***}p < 0.001$

an interaction term, a positive error process was operative in the model with an interaction term.

As Tables 4.17 and 4.18 show, in only one election in the Sixth American Party System was there a significant interaction effect. This was in 1976. The inclusion of the interaction term had little effect on inferences regarding the density and spatial terms in the model. The population density variable was still typically significant and positive. Only in the 1976 election did the effect become insignificant in the model with the interaction term where it was significant in the model without this term. Likewise, a positive spatial error process was operative in all of the same elections in the models with the interaction terms as it was in the models without.

Table 4.17 Spatial models of county-level voting in the sixth party system, 1968–1992

Year	ρ	λ	Constant	Proportion African American	Proportion Foreign Born	Proportion African American × Proportion Foreign Born	Population Density
1968		0.78***	−0.19***	0.79***	0.76***	1.52	0.01
		(0.01)	(0.01)	(0.04)	(0.21)	(1.14)	(0.01)
1972	0.77***		−0.11***	0.90***	0.47***	1.25	0.03*
	(0.01)		(0.01)	(0.02)	(0.11)	(0.90)	(0.01)
1976		0.81***	0.03	0.53***	−0.91***	3.41**	0.03
		(0.01)	(0.01)	(0.04)	(0.22)	(1.14)	(0.02)
1980		0.83***	−0.19***	0.88***	−0.34	1.73	0.06**
		(0.01)	(0.01)	(0.04)	(0.20)	(0.99)	(0.02)
1984		0.83***	−0.19***	0.86***	−0.30	1.56	0.06**
		(0.01)	(0.01)	(0.04)	(0.19)	(0.92)	(0.02)
1988		0.84***	−0.20***	1.01***	−0.10	0.72	0.08***
		(0.01)	(0.02)	(0.04)	(0.20)	(0.95)	(0.02)
1992		0.83***	−0.06***	0.86***	0.08	−0.20	0.08***
		(0.01)	(0.01)	(0.03)	(0.17)	(0.80)	(0.02)

$^{*}p < 0.05$; $^{**}p < 0.01$; $^{***}p < 0.001$

Table 4.18 Spatial models of county-level voting in the sixth party system, 1996–2016

Year	ρ	λ	Constant	Proportion African American	Proportion Foreign Born	Proportion African American × Proportion Foreign Born	Population Density
1996		0.81***	−0.07***	0.83***	0.50*	−0.68	0.04***
		(0.01)	(0.01)	(0.03)	(0.20)	(0.85)	(0.01)
2000		0.82***	−0.26***	0.96***	0.01***	−0.03	0.05***
		(0.01)	(0.01)	(0.04)	(0.00)	(0.79)	(0.01)
2004		0.87***	−0.33***	1.33***	0.77***	0.17	0.09***
		(0.01)	(0.02)	(0.04)	(0.09)	(0.58)	(0.02)
2008		0.88***	−0.30***	1.52***	1.07***	0.18	0.09***
		(0.01)	(0.02)	(0.04)	(0.09)	(0.56)	(0.02)
2012		0.89***	−0.37***	1.66***	1.06***	0.22	0.09***
		(0.01)	(0.02)	(0.04)	(0.08)	(0.52)	(0.02)
2016		0.85***	−0.52***	1.72***	1.65***	−0.79	0.13***
		(0.01)	(0.02)	(0.04)	(0.09)	(0.49)	(0.02)

$^{*}p < 0.05$; $^{**}p < 0.01$; $^{***}p < 0.001$

4.9 Conclusion

Chapter 2 identified the strong spatial structuring of county-level voting in the United States since 1828. In each election, counties have exhibited similar partisan voting as their neighbors. In this chapter we modeled this partisan voting using a set of common covariates for each election that tap major political faultlines of race, immigration, and the rural/urban divide.

We periodized our analysis using American party systems. Among our principal findings are that the sources of partisan voting change over time, and in understandable ways. For example, *Proportion African American* influenced partisan voting during both the Reconstruction Era and the Jim Crow Era, but in opposite directions. The size of local immigrant populations did not influence voting early in the country's history, but did as subsequent waves of immigrants raised the political salience of the immigration issue. In contrast to our current era, rural and urban counties didn't differ in their partisan voting in the nineteenth century. But by the 1920s, more densely populated counties became consistent supporters of the Democratic Party. Spatial diffusion processes were operative during the era of strong parties in the nineteenth century while spatial error processes have become dominant in recent decades while interpersonal connections between citizens have frayed.

References

Anselin, L. (1988). *Spatial econometrics: Methods and models*. Dordrecht: Kluwer.

Anselin, L. (2005). Spatial regression analysis in R: A workbook. Typescript. Center for Spatially Integrated Social Science, University of Illinois at Urbana-Champaign.

Anselin, L., & Rey, S. (1991). Properties of tests for spatial dependence in linear regression models. *Geographical Analysis, 23*(2), 112–131.

Anselin, L., Bera, A. K., Florax, R., & Yoon, M. J. (1996). Simple diagnostic tests for spatial 1634 dependence. *Regional Science and Urban Economics, 26*, 77–104.

Bera, A. K., & Yoon, M. J. (1993). Specification testing with locally misspecified alternatives. *1645 Econometric Theory, 9*, 649–658.

Brambor, T., Clark, W. R., & Golder, M. (2006). Understanding interaction models: Improving empirical analyses. *Political Analysis, 14*(1), 63–82.

Darmofal, D. (2015). *Spatial analysis for the social sciences*. New York: Cambridge University Press.

King, D. S., & Smith, R. M. (2005). Racial orders in American political development. *American Political Science Review, 99*, 75–92.

Mink, G. (1990). *Old labor and new immigrants in American political development: Union, party and state, 1875–1920*. Ithaca: Cornel University Press.

Putnam, R.D. (2000). *Bowling alone: The collapse and revival of American community*. New York: Simon & Schuster.

Robinson, W. S. (1950). Ecological correlations and the behavior of individuals. *American Sociological Review, 15*(3), 351–357.

Chapter 5
Clinton vs. Trump: Back to the Future?

> Jackson's victory [in 1828] shook the establishment like an earthquake. Henry Clay, Secretary of State for the defeated President John Quincy Adams, called Jackson's victory "mortifying and sickening." Oh, boy, does this sound familiar. Have we heard this? This is terrible. He said there had been "no greater calamity" in the nation's history.
>
> President Donald Trump at The Hermitage, Nashville, Tennessee, on the 250th Anniversary of the Birth of Andrew Jackson, March 15, 2017.[1]

The quote above is but one of many times that President Trump, since taking office, has compared himself to the country's 7th president, Andrew Jackson. The parallels between Jackson's election to the presidency in 1828 and Trump's in 2016 are indeed striking. Both Jackson and Trump ran as outsiders to the political establishment, populist heroes who would represent forgotten Americans whose concerns had been ignored by a corrupt Washington elite. Both ran for president after several decades of larger-than-life celebrity identities. Both were known for their antagonisms toward politicians and others whom they thought had disrespected them. And both defeated quintessential establishment politicians (John Quincy Adams in Jackson's case, Hillary Clinton in Trump's) who represented business as usual to both them and their supporters. Given these similarities, it is unsurprising that President Trump has placed a portrait of Andrew Jackson prominently in his Oval Office.

While the question of how similar Trump's presidency will be to Jackson's is one for future historians, we can examine whether their successful candidacies reflect a shared electoral base. Did counties that supported Jackson in 1828 tend to support

[1] https://www.whitehouse.gov/briefings-statements/remarks-president-250th-anniversary-birth-president-andrew-jackson/, retrieved on December 23, 2017.

D. Darmofal, R. Strickler, *Demography, Politics, and Partisan Polarization in the United States, 1828–2016*, Spatial Demography Book Series 2,
https://doi.org/10.1007/978-3-030-04001-7_5

Trump in 2016? This is a question we'll answer in this chapter. Before addressing this question, however, it is critical first to examine the Jacksonian tradition that Trump is said to reflect.

In his influential study of the Jacksonian tradition in American politics and American foreign policy, Mead (1999/2000, 9) highlights this populist tradition's core conception of America as a folk community "bound together by deep cultural and ethnic ties" in which some Americans are recognized as members of the community while others are perceived as outside the community. Mead also identifies five core principles of Jacksonian America that track with common popular conceptions of Trump's base of support among white working-class voters in the 2016 presidential election.

The first principle of the Jacksonian code identified by Mead is self-reliance. Americans who reflect the Jacksonian tradition pride themselves on making their own way in the world and, importantly, abjure dependence either on government via welfare *or* inherited wealth (12). This populist middle-way in which a Jacksonian middle-class equally rejects perceived pillars of the left and the right in American politics is a common theme of the Jacksonian tradition that sets it apart from the less populist Hamiltonian, Jeffersonian, and Wilsonian traditions with which Mead contrasts the Jacksonian tradition.

The second principle of Jacksonian America is equality. Importantly, this is defined as equality *within* the folk community. Among those citizens who are perceived as self-reliant and recognized as members of the American folk community, there is absolute equality based on the rights of the individual. This leads naturally to the third Jacksonian principle: individualism. As Mead (13) notes, "not just the right to self-fulfillment" but also "a duty to seek it." But as Mead (13) also notes, individualism in Jacksonian America must also follow a code of accepted mores including "loyalty to family, raising children "right", sexual decency... and honesty within the community." Individualism in the Jacksonian tradition is thus not carte blanche to behave as one wishes, but instead is tempered by a commitment to the tribal values of the folk community.

Mead defines the fourth principle of Jacksonian America as financial *esprit*, meaning a belief in buying on credit as both a right and an opportunity. As Mead notes, "Jacksonians have always supported loose monetary policy and looser bankruptcy laws." A corollary of this principle can be seen in modern Jacksonians' support for federal governmental programs such as Social Security, which are viewed not as welfare programs but instead as rights due to hard-working middle-class members of the folk community (Mead 1999/2000, 2016).

As Mead (2016) goes on to note, the definition of America as a "folk community" in which certain citizens are deemed valid members and others are not carries with it a dark side. In Mead's words, "Jacksonian sentiment embraces a concept of the United States as a folk community and, over time, that folk community was generally construed as whites only. Lynch law and Jim Crow were manifestations of Jacksonian communalism, and there are few examples of race, religious or ethnic prejudice in which Jacksonian America hasn't indulged."

At its core, as a non-ideological conception focused on a folk community of an in-group vs. out-groups, the Jacksonian tradition carries a strong element of identity politics, that is "political activity or movements based on or catering to the cultural, ethnic, gender, racial, religious, or social interests that characterize a group identity" (Random House 2018). This is important for our understanding of the 2016 election given the evidence scholars have amassed that identity politics (along with declining health trends) played a greater role in Trump's Electoral College victory than objective economic distress.

Since the surprising outcome of the 2016 presidential election, countless stories have been written on and countless interviews conducted of white working-class Trump supporters in key swing states in the Midwest. At their foundation, these stories by journalists are aimed at explaining the election's surprising result. But while many of these stories focus on individual cases of economic hardship leading white working-class voters to favor Trump, the evidence of several academic studies of the 2016 election points us in a different direction. Rather than objective economic distress leading to Trump votes, particularly among the white working class, these studies point instead to the importance of identity politics, particularly white identity politics, in producing the surprising outcome on election night 2016.

Indeed, at the macro level, the evidence is not one of economic distress. Instead, on election night 2016, there had been 73 consecutive months of positive job growth following the Great Recession. True, wages were stagnant, but this was not a new phenomenon that began during the Obama presidency; wages have instead been stagnant since 1973 (Cowen 2015). As Sides et al. (2016, 54–56) show, unemployment had declined markedly from 2008 to 2015 and the index of consumer sentiment—a combination of current evaluations and future expectations regarding the economy—was as high in 2015 as it had been at the beginning of Ronald Reagan's "Morning in America" re-election campaign in 1983.

Indeed, Sides et al. (2016) make the important point that Hillary Clinton's popular vote victory is what we would have expected from a forecasting model based on economic performance and incumbent presidential approval. The anomaly was Trump's Electoral College victory. Two factors were critical in producing this electoral victory. The first, and most important, was partisan identification, the single biggest predictor of voting behavior in the United States. With a closely divided electorate in terms of partisan loyalties, and partisanship increasingly shaping perceptions of the economy (and depressing Republicans' evaluations of the improving economy), 2016, like other recent elections, was primed to be a close election (Sides et al. 2016, 52).

The second important factor producing Trump's Electoral College win was the "racialization" of politics in the preceding decades, and particularly during the preceding 8 years of the Obama presidency. As Sides et al. (2016, 52) note, "white Americans' opinions of blacks and other minority groups became more intertwined with their partisan identities and more potent predictors of their opinions about the economy and many other issues—virtually anything even associated with Obama. This growing "racialization" of public opinion began before Obama became president, but it became more pronounced during his time in office."

With whites increasingly identifying with the Republican Party and racial and ethnic minorities increasingly identifying with the Democrats—and racial attitudes increasingly predicting public opinion—racial and other demographic factors were primed to be central to the 2016 election. As Sides et al. (2016, 64) state, "Even without Barack Obama on the ballot in 2016, trends in previous decades—trends that only accelerated when Obama was president—made issues concerning race, ethnicity, and religion central to the strategies of the 2016 presidential candidates and immanent in the election-year discourse." In short, identity politics was more consequential to the 2016 campaign and to the 2016 election outcome than a simple model built on fundamentals such as objective economic indicators would have predicted.

Sides et al.'s (2018) results are also consistent with the idea that Trump tapped into a longstanding Jacksonian tradition, although they do not connect their analysis to this tradition. However, they find that Trump "succeeded by activating longstanding, but often unappreciated, sentiments among Republican voters about immigration and economic entitlements" (Sides et al. 2018, 135). And, specifically, they find that Trump's candidacy attracted supporters who had preexisting opposition to immigration and support for entitlement programs such as Social Security and Medicare. As noted above, both sets of policy preferences are consistent with the Jacksonian tradition.

Other studies have also highlighted the importance of identity politics in the 2016 election. Utilizing a survey conducted by the Public Religion Research Institute (PRRI) in conjunction with *The Atlantic*, Cox, Lienesch, and Jones find that economic hardship did not lead white working-class voters to vote for Trump in 2016. In fact, white working-class voters who stated that they were in either fair or poor financial shape were actually more likely to vote for Hillary Clinton over Donald Trump (at a $p < .1$ level). Instead, four factors led white working-class voters to vote for Trump: identification with the Republican Party, fears about cultural displacement, support for deporting immigrants who were living in the country illegally, and economic fatalism (Cox et al. 2017).

The partisan identification finding is consistent with Sides, Tesler, and Vavreck's analysis and is, again, what we should expect in a country in which party id is the single best predictor of voting at the individual level. Cox, Lienesch, and Jones find that white working-class voters who identified with the Republican Party were 11 times more likely to vote for Trump than white working-class voters who did not identify as Republicans. But the 2016 election was not simply a partisan identification election. Instead, three factors connected to white identity also played critical roles in the election's outcome.

The first of these was a fear of cultural displacement. Cox, Lienesch, and Jones find that white working-class voters who reported feeling "like a stranger in their own land and who believe the U.S. needs protecting against foreign influence were 3.5 times more likely to favor Trump than those who did not share these concerns." These beliefs are consistent with both white identity politics and in-group perceptions of the "folk community" that are intrinsic to the Jacksonian tradition in American politics.

Also reflective of the identity politics of the Jacksonian folk community, attitudes toward undocumented immigrants also shaped white working-class voting. Such voters who favored deporting undocumented immigrants were 3.3 times more likely to vote for Trump than those voters who did not favor deportation.

Economic fatalism also played a role in shaping white working-class voting preferences. A college education has long been seen as a key toward financial improvement. However, white working-class voters who viewed a college education as a gamble instead of an important investment in the future were nearly twice as likely to favor Trump over Clinton (Cox et al. 2017).

Cox, Lienesch, and Jones' findings regarding identity politics are reinforced by a separate study of whites during the 2016 primaries by Major et al. (2016). These authors find that ethnic (white) identification was a critical identification predicting support for Trump and for anti-immigrant policies as well as opposition to political correctness. Specifically, employing an experimental design, Major, Blodorn, and Blascovich find that reminding white Americans with high levels of white identification that whites will be a minority in the U.S. by 2042 leads such high ethnic identifiers to become more supportive of Trump and his immigration policies and less supportive of political correctness norms. Increased group status threat mediated these effects among these respondents. In contrast, among whites low in white identification, the reminder of the country's growing diversity did not produce perceptions of group status threat. Instead, such a reminder for whites low in white identification actually led them to have more negative attitudes toward Trump and reduced opposition to political correctness norms.

Trump ran strongest in rural locations, winning 61% of the vote in rural America to Clinton's 34% according to exit polls (in comparison, Clinton beat Trump in more populous urban America by a similar 60% to 34% margin).[2] Although Cramer's (2016) analysis was focused on the growing urban-rural divide in Scott Walker's Wisconsin, it is informative for the role that identity politics may have played in producing this geographic divergence in the 2016 election. Specifically, Cramer's research suggests that the rural voting pattern in the 2016 election may reflect a rural identity based upon resentment of urban America, which many rural residents feel gets both the lion's share of government attention and is in the cultural driver's seat in contemporary America.

Luttig et al. (2017) employ an experimental design and also find a strong racial component to support for Trump. These authors find that experimental subjects who supported Trump in 2016 were more likely when randomly exposed to a black man (as opposed to a white man) facing foreclosure to oppose housing assistance, be angry about such assistance, and to blame beneficiaries of such programs for their own situation. Subjects opposed to Trump exhibited the opposite effects (Luttig et al. 2017, 1). Clearly, racial cues affected Trump supporters and those opposing Trump differently during the 2016 election.

---

[2]https://www.cnn.com/election/2016/results/exit-polls, accessed March 15, 2018.

The role of race is reinforced by Tolbert et al.'s (2018) analysis. They find that racial resentment interacted with enthusiasm for Trump to produce support for Trump in the 2016 election. As Tolbert et al. (2018, 260) state, "Voters with prior racially resentful attitudes were much more likely to support Trump, but those who also felt enthusiasm generated by the Trump campaign were even more likely to rate him highly in both the Iowa caucus and general election campaigns, with racial resentment playing an even bigger role in driving evaluation of the candidate."

Donovan and Redlawsk (2018) place Trump's support in comparative perspective and find that it has more in common with support for right-wing populists in Australia, Canada, New Zealand, and the U.K. than it does for standard right-of-center candidates. Specifically, these authors find that support for Trump, like the support for other right-populist candidates, is based on racial resentment, restrictive immigration attitudes, and anxiety (Donovan and Redlawsk 2018, 190). The factors predicting support for Trump, in other words, depart from those predicting support for standard center-right Republican candidates in the U.S. and conform with those predicting support for right-populists in other countries.

Focusing their analysis on white working-class men, Turney et al. (2017) argue that the support for Trump among this group of voters was a culmination of several decades of movement to the right by this group. Echoing what others have found, these authors find that cultural factors predicted Trump support among white working-class men in 2016 more than economic concerns. Specifically, both attitudes toward immigration and racial resentment predicted Trump support more than economic attitudes did among white working-class male voters in 2016.

Federico and Golec de Zavala (2018, 1) find that a collective narcissism, defined as "an inflated, unrealistic view of the national ingroup's greatness contingent on external recognition", focused on America's national identity predicted support for Trump in 2016. Trump ran the most nationalist presidential campaign in decades, one that tapped into the Jacksonian tradition in promising to "Make America Great Again." Federico and Golec de Zavala argue that collective narcissism promotes perceptions of threat to the in-group as well as retaliatory hostility toward out-group members and that this is the mechanism by which nationalist identity promoted support for Trump.

Gender attitudes also played a critical role in a contest involving the first female major party presidential nominee in Hillary Clinton. Ratliff et al. (2017) examine the roles of two forms of sexism in the 2016 election. Trump voters had higher levels of hostile sexism, defined as "antagonistic attitudes toward women" as well as benevolent sexism defined as paternalistic attitudes toward women (1–2). The former produced more positive views of Trump and less positive views of Clinton and also predicted vote choice for Trump over Clinton. Importantly, this effect holds even controlling for ideology. The latter did not have an independent effect on vote choice after controlling for the effects of ideology and hostile sexism but was more common among Trump supporters than among Clinton supporters (1). Bracic et al.'s (2018) findings are largely consistent with Ratliff et al.'s. They find that paternalistic beliefs that men should have control over their wives had no independent effect on vote choice in the 2016 election. In contrast, beliefs that men are more emotionally suited for politics than women did produce more favorable attitudes toward Trump and made voters more likely to vote for him. Importantly, they find that these effects were most pronounced among white voters.

Other scholars have also found that objective economic factors were not the principal causes of Trump's support. In an extensive analysis of more than 125,000 Gallup survey respondents from July, 2015 through early October, 2016, Rothwell and Diego-Rosell find that while Trump's supporters were less educated and more likely to have blue-collar jobs than non-Trump supporters, they also had higher mean household incomes than the non-Trump supporters. Importantly, however, features of the communities in which Trump supporters live do predict their support. Specifically, as Rothwell and Diego-Rosell (2016, 1) note, "living in racially isolated communities with worse health outcomes, lower social mobility, less social capital, greater reliance on social security income and less reliance on capital income, predicts higher levels of Trump support."

Manza and Crowley (2017) examine Trump supporters during the Republican nomination process and find similar results regarding the social basis of Trump's support during that phase of the campaign. Contrary to journalistic accounts, Trump's support wasn't drawn from downscale voters in objective economic distress. Instead, Trump's supporters tended to have higher median incomes than the average voter in their states. They also tended to be better educated than the average voter (Manza and Crowley 2017, 5). In fact, there's evidence that even opinions on one of the presumed, central economic concerns of the 2016 election—trade policy—are themselves racially tinged, with racial attitudes predicting support for protectionism (Guisinger 2017).

While objective economic distress did not predict Trump support in 2016, negative health outcomes in voters' communities did. Rothwell and Diego-Rosell had found that negative health outcomes in communities predicted Trump support and Bilal et al. (2018) corroborate this result in their analysis of the 2016 election. Building on Case and Deaton's (2015) finding of increased mortality rates for middle-age non-Hispanic whites in the U.S., Bilal, Knapp, and Cooper find that such increased mortality predicted county-level vote shifts toward Trump in 2016. Specifically, they find that counties that flipped from supporting Obama in 2008 and 2012 to supporting Trump in 2016 had increases in mortality rates among middle-age non-Hispanic whites between 1999–2005 and 2009–2015 in comparison to counties that voted for Clinton in 2016. They also find that counties with greater health inequalities in life expectancy were more likely to vote for Trump in 2016 regardless of how they had voted in previous elections (Bilal et al. 2018, 33). Similarly, Bor (2017) finds that negative health outcomes were associated with increased county-level support for the Republican presidential candidacy of Donald Trump, with counties in which life expectancy declined or stagnated between 1980 and 2014 experiencing a 10 point shift in voting toward the Republican Party between 2008 and 2016. In short, while objective indicators did predict Trump support in 2016, they were health outcome indicators, not economic indicators. Trump's victory in the 2016 election appears to have been one driven by identity politics associated with the Jacksonian tradition and declining health outcomes, rather than economic hardship.

Both Andrew Jackson and Donald Trump presented themselves to the public as bold leaders unafraid of others' opinions of them. And both raised fears among political contemporaries regarding their potential for authoritarian tendencies. It is not surprising, therefore, that scholars have examined the effects of authoritarian attitudes in promoting support for Trump in 2016.

Choma and Hanoch (2017) find that two ideological beliefs were instrumental in shaping support for Trump in 2016. Right-wing authoritarianism is an ideological orientation in which individuals high in right-wing authoritarianism believe that "people should obey and respect authorities deemed as legitimate, abide by social conventions, and endorse harsh punishment of norm violators" (Choma and Hanoch 2017, 287). Social dominance orientation is an ideological belief that "relations between social groups should reflect a hierarchy with some groups wielding more power than others" (Choma and Hanoch 2017, 287). These authors find that right-wing authoritarianism and social dominance orientation predicted both more positive attitudes toward Trump and vote preference for Trump in 2016.

5.1 The Relationship Between the 1828 and 2016 Elections

A critical question regarding the 1828 Jackson vs. Adams contest and the 2016 Trump vs. Clinton contest is whether the political geography of these two elections was similar. Put differently, were Andrew Jackson and Donald Trump's electoral bases similar? Did they exhibit electoral strength in the same counties? Or are the political geographies of the two elections unrelated to each other?

One way to examine this is to examine the Pearson product-moment correlation coefficient for the county-level margin of victory variable in these two elections. A reminder from Chap. 2 that the margin variable is a continuous variable measuring the Democratic proportion of the vote minus the National Republican, Whig, or Republican proportion of the vote (depending upon the era). In this case, for the 1828 presidential election, it is the proportion of the vote in the county for the Democratic candidate for president (Andrew Jackson) minus the proportion of the vote in the county for the National Republican candidate for president (John Quincy Adams). And for the 2016 presidential election, it is the proportion of the vote in the county for the Democratic candidate (Hillary Clinton) minus the proportion of the vote in the county for the Republican candidate (Donald Trump).

The Pearson's correlation coefficient for the margin variable in these two elections was −.270. This reflects a modest negative relationship in the partisan vote in these two elections. But, importantly, a strongly significant one. The correlation coefficient is significant (in a two-tailed test) at a $p < .001$ level. In other words, support for Andrew Jackson in a county in 1828 was significantly associated with support for Donald Trump in the county 188 years later.

Table 5.1 presents the Pearson's correlation coefficients with 2016 for all presidential elections as well as whether these correlations were significant at a $p < .001$ level. As we can see, voting in the 1828 election was more similar to

Table 5.1 Correlation with 2016 voting

Election	Correlation coeff.	Election	Correlation coeff.
1828	−0.270*	1920	0.078*
1832	−0.184*	1924	0.008
1836	−0.158*	1928	0.200*
1840	−0.157*	1932	0.001
1844	−0.192*	1936	0.125*
1848	−0.267*	1940	0.144*
1852	−0.235*	1944	0.166*
1856	−0.278*	1948	−0.028
1860	−0.242*	1952	0.095*
1864	0.046	1956	0.062*
1868	−0.014	1960	0.195*
1872	−0.256*	1964	0.052
1876	−0.196*	1968	0.388*
1880	−0.185*	1972	0.453*
1884	−0.154*	1976	0.085*
1888	−0.084*	1980	0.352*
1892	0.072*	1984	0.518*
1896	−0.112*	1988	0.455*
1900	0.023	1992	0.499*
1904	0.078*	1996	0.605*
1908	0.042	2000	0.766*
1912	0.117*	2004	
1916	0.054		

*Significant at $p < 0.001$

voting in the 2016 election than all other nineteenth century elections were, with the one exception of the 1856 contest between James Buchanan and John C. Frémont. In fact, county-level results in the 1828 election were more strongly correlated with the 2016 election results than any election until the 1968 election, with the lone exception of the 1856 contest.

Most nineteenth century elections do exhibit a correlation with the 2016 election and where this correlation exists, the sign is negative. In other words, where Democrats ran strong in the nineteenth century, Trump ran strong in the 2016 election. This is as we would expect for a party forged in the Jacksonian tradition.

While significant negative correlations between nineteenth century elections and the 2016 election were common, this pattern disappeared in the Fourth Party System. In fact, 1896 was the last election in which county-level voting was negatively correlated with voting in the 2016 election. Following this, every other presidential election for decades was uncorrelated with voting in the 2016 election while in the other elections, the correlation was a positive one. Note, however, that even though these elections in the Fourth Party System were much closer temporally to the 2016 election, none had as strong a correlation with voting in that election as did the 1828 election.

As we move into the Fifth Party System, we see that the correlations with the 2016 election become more consistently significant. Counties with strong Democratic (Republican) support in the New Deal party system tended to show strong support for Clinton (Trump) in 2016. And yet, still, none of these positive correlations during this party system are as strong as the negative correlation between the 1828 and 2016 elections.

Correlations with the 2016 election become markedly stronger beginning in 1968. This lends empirical support to our argument that we have entered a distinct Sixth American Party System. Indeed, these correlations in all elections but 1976 are much larger than the correlation with the 1828 election. In short, it took until the past few decades for more immediate elections to exhibit a stronger relationship with 2016 than the 1828 contest had. Note, again, however, the small correlation between the anomalous 1976 election and 2016 (less than a third of the correlation between 1828 and 2016). This highlights once again the problem of Bishop choosing an anomalous election as a baseline for his Big Sort argument.

5.2 Looking Further on the 1828 and 2016 Relationship

We can understand the relationship between Jackson's support in 1828 and Trump's in 2016 by delving further into county-level voting in these two elections. First, let's look at how counties that were particularly strong Jackson counties in 1828 voted in 2016. As a first cut at this, we'll look at counties in the top 25% of the distribution of support for Jackson in 1828. These are the counties that had margins of victory for Jackson over Adams of 60 percentage points or more.

These counties had an average margin of victory for Trump in 2016 of 37.0 percentage points. In contrast, counties in the lower 75% of the 1828 distribution had average Trump margins of only 25.1 percentage points. The difference in 2016 margins of 11.9 percentage points for the pro-Jackson vs. less pro-Jackson counties in 1828 is statistically significant at a $p < .0001$ level whether one uses a one- or two-tailed test.

We can alter the comparison by looking at the ten percent most pro-Jackson counties in 1828 vs. the rest of the counties from that election. The top ten percent of the distribution in 1828 is defined as those counties that had pro-Jackson margins of 82 percentage points or more. These counties had pro-Trump margins in 2016 that averaged 43.3 percentage points. In contrast, the bottom ninety percent of the 1828 distribution had an average pro-Trump margin of 26.3 points. The strongest pro-Jackson counties in 1828 on average supported Trump by 17 percentage points more than the other 90% of the 1828 counties. Once again, this difference is statistically significant at a $p < .0001$ level whether one uses a one- or two-tailed test.

County-level support for Andrew Jackson in 1828 predicted support for Donald Trump 188 years later. Now let's examine a much more recent election—Bishop's baseline election of 1976. Did county-level voting in 1976 predict voting 40 years later to the same extent that voting in 1828 did 188 years later?

We again look at counties in the top 25% of the margin of victory distribution. In this case, this is counties with Carter margins of victory of 20 percentage points or more. These strong Carter counties in 1976 had an average Trump margin of victory of 31.5 percentage points. In contrast, the counties in the lower 75% of the 1976 distribution had an average Trump margin of victory of 32.3 percentage points. In short, there was only a minor difference in voting in 2016 of 0.8 points between the strong pro-Carter counties and the rest of the 1976 counties. This difference is not statistically significant.

Perhaps more of a predictive effect of 1976 voting on 2016 voting exists if we look at the most strongly pro-Carter counties from 1976. The 10% of counties that were most pro-Carter in 1976 had an average margin of victory in that election of 40 percentage points or more. These most strongly pro-Carter counties from 1976 had average margins of victory for Trump in 2016 of 32.5 percentage points. This was actually slightly more pro-Trump than the rest of the 1976 distribution, which averaged a 32.1 point Trump margin, although this difference of 0.4 percentage points is not statistically significant.

In summary, the Jackson-Adams election of 1828 predicted voting in the 2016 presidential election 188 years later much better than the Carter-Ford election of a mere 40 years prior to 2016. The 2016 election appears to some extent to reflect a Jacksonian tradition dating back to the early nineteenth century.

5.3 Trump and Jackson: A Closer Look

We can get more insights into the support for Jackson and Trump, and for Adams and Clinton, by looking at two counties that changed their allegiances over the course of America's electoral history. Pike County, Kentucky is a classic Appalachian county in Eastern Kentucky that was one of Andrew Jackson's strongest counties in 1828 and one of Donald Trump's strongest in 2016. Hampshire County, Massachusetts tells the opposite story. A New England county featuring several higher education institutions, it went from being a strong county for John Quincy Adams in 1828 to being a strong county for Hillary Clinton in 2016.

Pike County, Kentucky is an Appalachian county famous in American history as the site of the decades-long feud between the Hatfields and the McCoys (Ban 2014). A county in which coal mining has long been a central part of its local economy, Pike County is also a strong Trump county. In 2016, it gave slightly more than 80% of its votes to the Republican presidential candidate.

The citizens of Pike County reflect Jacksonian values. Consistent with the tradition of gun ownership in Jacksonian portions of America, the Pike County School Board became one of the first in the country to move to allow teachers to carry concealed weapons in schools following the shootings at Marjory Stoneman Douglas High School. Reflecting the strong gun culture in this part of the country, the school board's vote was unanimous (Wright 2018).

But while Pike County is a strong Republican county in presidential elections today, this wasn't always the case. In 1828, Pike County favored Andrew Jackson over John Quincy Adams by 97 percentage points. In other words, Pike County had a partisan shift between 1828 and 2016 of an astounding 177 percentage points. And yet while Jackson's and Trump's party labels differed, this Appalachian county in the heart of the Jacksonian stronghold in rural America appears to have found an affinity with these two quite similar presidential candidates.

Figure 5.1 presents the partisan vote in Pike County from 1828 to the present. As can be seen from this figure, Pike County has an interesting partisan history. Finding a champion for its values and its way of life, it went strongly for Jackson in both the 1828 and 1832 elections and its margin for Jackson's Vice President, Martin Van Buren, was actually slightly stronger in 1836 than it had been for Jackson in 1832. But perhaps feeling the effects of the Panic of 1837 and ensuing recession, it shifted strongly toward Whig William Henry Harrison in 1840. It would maintain its Whig allegiance for the three subsequent elections, though tellingly, giving its smallest Whig margin to fellow Kentuckian Henry Clay in his contest with the Jacksonian James K. Polk in 1844.

In 1856, faced with a choice between Democrat James Buchanan and the candidate of the avowedly anti-slavery Republican Party, John C. Frémont, Pike County turned sharply toward the Democrats. Pike would continue to display this Democratic allegiance throughout the Third Party System. This changed, however, with the advent of the Fourth Party System in 1896. Like much of the rest of the country, Pike County shifted toward the Republicans. It would, in fact, vote for the Republican candidate for president in every election during the Fourth Party System.

The Great Depression and subsequent New Deal policies of Franklin Delano Roosevelt led Pike County to once again shift its loyalties to the Democratic Party. FDR carried the county by more than 20 points in 1932 and Pike would continue to support Democratic presidential candidates by double-digit margins in subsequent elections in the Fifth Party System. The lone exception to Democratic victories in the county during the New Deal party system was Eisenhower's 1956 re-election. The strength of the Democratic loyalties in Pike County is reflected in the fact that this county in heavily Protestant Kentucky went for the Catholic candidate, John F. Kennedy, by 13 percentage points. In the last election of the Fifth Party System, Pike favored Democrat Lyndon Baines Johnson over Barry Goldwater by 33 percentage points.

While extensive portions of the country shifted their allegiance to the Republicans at the presidential level with the advent of the Sixth Party System in 1968, this wasn't the case in Pike. Pike favored Hubert Humphrey over Richard Nixon in 1968 by 12 percentage points. Although Nixon did win Pike County handily, by 14 points, in his 1972 contest with George McGovern, this was an anomaly during this period. Pike continued to favor Democrats, voting for Jimmy Carter over Gerald Ford in 1976, Carter over Ronald Reagan in the latter's landslide victory in 1980, Walter Mondale over Reagan in the latter's even larger 1984 landslide, Michael Dukakis by 24 percentage points over George H.W. Bush in 1988, Bill Clinton by 33 points

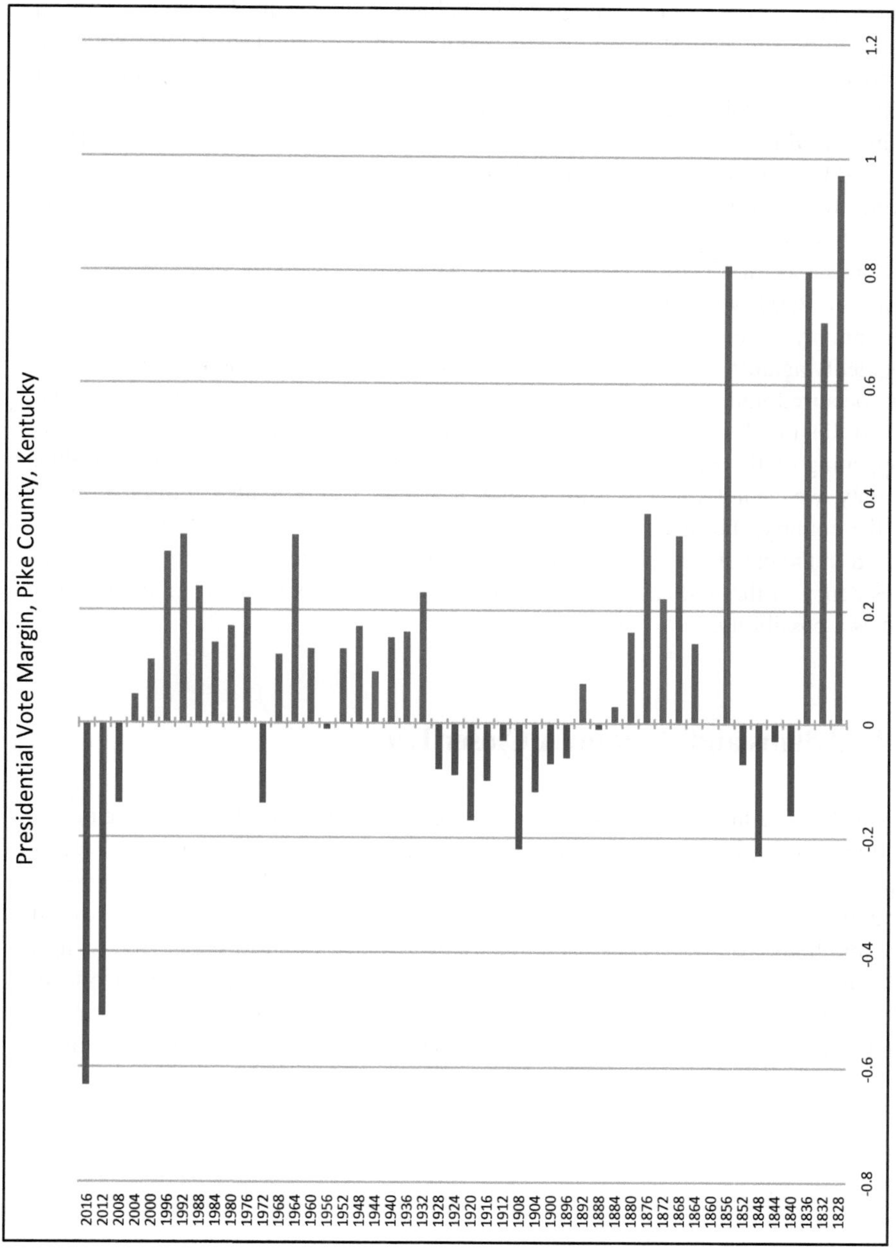

Fig. 5.1 Presidential voting in Pike County, Kentucky

over Bush in 1992 and by 30 points over Bob Dole in 1996, Gore over George W. Bush in 2000, and John Kerry over Bush in 2004.

Things changed markedly, however, with the candidacy of the first African-American major party presidential nominee, Barack Obama, in 2008. While Obama was getting the largest share of the popular vote of any Democratic nominee nationally since LBJ in 1964, Pike County shifted sharply to the Republicans in 2008, favoring John McCain over Obama by nearly 14 percentage points. In 2012, it favored Mitt Romney over Obama by nearly 51 percentage points. And in 2016, it favored Donald Trump over Hillary Clinton by nearly 63 percentage points. In other words, between Bill Clinton's election in 1992 and the candidacy of Hillary Clinton in 2016, Pike County shifted to the Republican Party by 96 percentage points.

The break in the time series is clear. The shift toward the Republicans happened in 2008, with Obama's candidacy. A county that had favored Democrats, presumably on economic grounds, since the beginning of the New Deal, shifted decidedly against the Democratic nominee during a sharp economic downturn that had occurred under a Republican president. What was different about Obama that led to this rapid shift toward the Republicans in 2008 during a recession that was happening on the Republican watch? Lacking individual-level survey data on Pike County, we cannot know for sure. However, if Pike is representative of other areas of the country, the racialization of political attitudes that political scientists have traced to the candidacy of the first African-American major party nominee and the presidency of the first African-American president Barack Obama cannot be ruled out as a possibility.

5.4 Adams and Clinton: A Closer Look

While Andrew Jackson was racking up an enormous margin of 97 percentage points in Pike County, Kentucky, President John Quincy Adams was racking up an almost equally enormous margin of 87 percentage points over 600 miles away in Hampshire County, Massachusetts. Later in the nineteenth century, Hampshire County would become home to Emily Dickinson, one of America's foremost poets. Today, it is known as the home of the "Five Colleges": Amherst College, Hampshire College, Mount Holyoke College, Smith College, and the University of Massachusetts Amherst. And with a such a strong educational base, Hampshire bills itself on its website as "a culturally rich, diverse area chock full of intellectualism, innovation, and ivy"[3] and as "one of those pockets on the planet where you can be yourself—any side, any shade, any shape".[4] With an ethos of diversity, education, and innovation and with nearly 45% of its population 25 years and over having a Bachelor's degree or higher,[5] Hampshire County, Massachusetts is now a quintessentially blue county. In 2016, Hillary Clinton carried Hampshire County by 40 percentage points, reflecting a 127 percentage point shift from the support for the National Republican Adams in 1828.

Figure 5.2 presents the partisan vote in Hampshire County from 1828 to the present. Hampshire County's partisan history is a common one for many counties in New England. The county was first a stronghold of support for the National Republican candidacies of Adams in 1828 and Clay in 1832. With the emergence of the Whig Party by 1836, it became a stronghold for the Whigs throughout the remainder of the Second Party System. From 1836 to 1852, the Whigs' margin over the Democrats in Hampshire County ranged from 31 to 42 points.

The Third Party System began with a large increase in the vote against the Democratic Party. The shift from the Whigs to the Republicans as the main competition for the Democrats produced an increase in the presidential vote margin from 31 points in 1852 to 69 points in 1856. These large Republican margins would continue in the 1860, 1864, and 1868 elections, with margins of 71, 71, and 73 points, respectively. The Republican margins would henceforth generally decline over the remainder of the Third Party System. The 73 point margin in 1868 was followed up by a 50 point margin in 1872 and then a 33 point margin in 1876. There would be a slight uptick in Republican support in 1880, with a 40 point margin. In the remaining three elections in the Third Party System, the Republican margins would be a much reduced 17, 16, and 14 points. Still, there was no election in the Third Party System in which the Republicans enjoyed less than a double digit advantage in Hampshire County. The county was clearly a solid Republican county from the beginning of the party, following on the several decades of preceding support for the Whigs and the National Republicans.

The Fourth Party System kicked the Republican margins in Hampshire County up again. The 14 point margin in 1892 swelled to a 57 point margin in 1896. The increased Republican margins continued for the next three elections. Although reduced in the three-party election of 1912 and in Wilson's re-election in 1916, the Republican margins bounced back up in the Republican landslide elections of 1920

[3]https://visithampshirecounty.com/explore/culture/the-five-colleges, accessed on March 3, 2018.

[4]https://visithampshirecounty.com/, accessed on March 3, 2018.

[5]https://factfinder.census.gov/faces/tableservices/jsf/pages/productview.xhtml?src=CF, accessed on March 3, 2018.

and 1924. Importantly, however, the margin was much reduced in 1928 in the final election of the Fourth Party System. Hoover's margin over Al Smith would be only 5 percentage points and this would be a harbinger of things to come in the Fifth Party System.

Hoover still outpaced Franklin Delano Roosevelt in Hampshire County in the first election of the Fifth Party System in 1932. This would, however, be the Republican high-water mark for two decades. FDR won Hampshire by 5 points in 1936, 6 points in 1940, and 8 points in 1944. Truman carried the county by a more modest 2 percentage points in 1948. Eisenhower carried Hampshire by larger margins of 17 and 24 points in his two election wins. The 1956 election would, however, be the last time a Republican presidential candidate carried Hampshire County to date.

In 1960, Hampshire voted for fellow Bay Stater John F. Kennedy by 14 points. Lyndon Johnson's margin over Goldwater in 1964 was a whopping 47 percentage points. Humphrey defeated Nixon in Hampshire by 23 points and even George McGovern defeated Nixon by 8 points in 1972 (in the only state he would carry in that election). Carter carried Hampshire County by double digits in both of his elections and Mondale carried it by 12 points while losing nationally to Reagan by more than 18 points.

Beginning with Dukakis's victory in Hampshire in 1988 by 24 points, no Democratic candidate to date has carried the county by less than 20 points. Bill Clinton's two victories were by 32 points in 1992 and by 41 points in 1996. Gore's margin in 2000 was a somewhat reduced but still robust 28 points.

The 2004 election, one with strong cultural overtones, represents another step up in Democratic support in Hampshire. John Kerry bested incumbent President George W. Bush by 41 points in 2004. Obama's margins of victory were 46 points in 2008 and 44 points in 2012. As stated earlier, in 2016 Hillary Clinton beat Donald Trump by 40 points in Hampshire County.

Looking again at Fig. 5.2, we can see what a marked partisan change Hampshire County has undergone since the advent of mass voter participation in 1828. In that election, the county voted overwhelmingly for a candidate with considerable experience in government, a former Senator and Secretary of State with a personal connection to a former president. In 2016 it would do exactly the same thing. But in 1828 it was voting for a National Republican and in 2016 it was voting for a Democrat. The Republicans, the dominant party of the Second, Third, and Fourth Party Systems in Hampshire County, lost their advantage in the Fifth Party System and have lost it further since the Kerry-Bush election of 2004. What was once a stronghold of Republican support is now a stronghold of Democratic support.

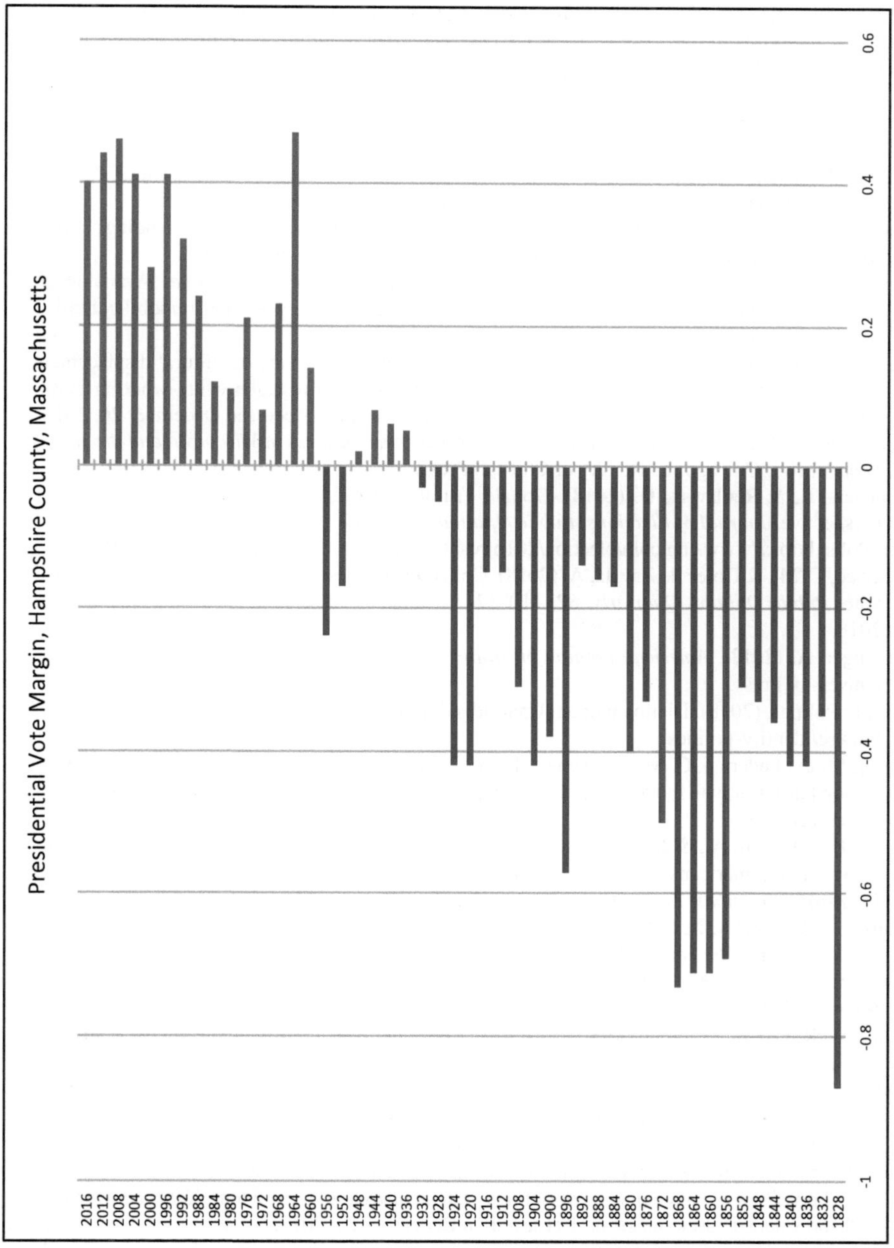

Fig. 5.2 Presidential voting in Hampshire County, Massachusetts

References

Ban, C. (2014). County innovations and solutions. National Association of Counties. http://www.naco.org/articles/county-innovations-and-solutions-6. Retrieved on March 2, 2018.

Bilal, U., Knapp, E. A., & Cooper, R. S. (2018). Swing voting in the 2016 presidential election in counties where midlife mortality has been rising in white non-Hispanic americans. *Social Science & Medicine, 197*, 33–38.

Bor, J. (2017). Diverging life expectancies and voting patterns in the 2016 US presidential election. *American Journal of Public Health, 107*, 1560–1562.

Bracic, A., Israel-Trummel, M., & Shortle, A. (2018) Is sexism for white people? Gender stereotypes, race, and the 2016 presidential election. *Political Behavior*

Choma, B. L., & Hanoch, Y. (2017). Cognitive ability and authoritarianism: Understanding support for Trump and Clinton. *Personality and Individual Differences, 106*, 287–291.

Cowen, T. (2015). It's not the inequality; it's the immobility. *New York Times* Available at: https://www.nytimes.com/2015/04/05/upshot/its-not-the-inequality-its-the-immobility.html. Accessed March 4, 2018.

Cox, D., Lienesch, R., & Jones, R. P. (2017). Beyond economics: Fears of cultural displacement pushed the white working class to Trump. https://www.prri.org/research/white-working-class-attitudes-economy-trade-immigration-election-donald-trump/. Accessed December 26, 2017.

Cramer, K. J. (2016). *The politics of resentment: Rural consciousness in Wisconsin and the rise of Scott Walker.* Chicago: University of Chicago Press.

Donovan, T., & Redlawsk, D. (2018). Donald Trump and right-wing populists in comparative perspective. *Journal of Elections, Public Opinion and Parties, 28*(2), 190–207.

Exit Polls. https://www.cnn.com/election/2016/results/exit-polls. Accessed March 15, 2018.

Federico, C. M., & Golec de Zavala, A. (2018). Collective narcissism and the 2016 US presidential vote. *Public Opinion Quarterly, 82*, 110–121. Available online first. Accessed February 25, 2018.

Guisinger, A. (2017). *American opinion on trade: Preferences without politics*. Oxford: Oxford University Press.

Identity Politics. (2018). Dictionary.com Unabridged. Random House. http://www.dictionary.com/browse/identity-politics

Luttig, M. D., Federico, C. M., & Lavine, H. (2017). Supporters and opponents of Donald Trump respond differently to racial cues: An experimental analysis. *Research and Politics, 4*. October–December, 1–8.

Major, B., Blodorn, A., & Major Blascovich, G. (2016). The threat of increasing diversity: Why many white Americans support Trump in the 2016 presidential election. *Group Processes & Intergroup Relations, 21*, 1–10. Available online first. Accessed March 1, 2018.

Manza, J., & Crowley, N. (2017). Working class hero? Interrogating the social bases of the rise of Donald Trump. *The Forum, 15*(1), 3–28.

Mead, W. R. (1999/2000). The Jacksonian tradition and american foreign policy. *The National Interest, 58*: 5–29.

Mead, W. R. (2016). Andrew Jackson, revenant. *The American Interest*. Accessed on March 20, 2018.

Ratliff, K. A., Redford, L., Conway, J., & Smith, C. T. (2017). Engendering support: Hostile sexism predicts voting for Donald Trump over Hillary Clinton in the 2016 U.S. presidential election. *Group Processes & Intergroup Relations*, 1–16.

Remarks by the president on 250th anniversary of the birth of President Andrew Jackson. https://www.whitehouse.gov/briefings-statements/remarks-president-250th-anniversary-birth-president-andrew-jackson/. Retrieved on December 23, 2017.

Rothwell, J., & Diego-Rosell, P. (2016). Explaining nationalist political views: The case of Donald Trump. Manuscript.

Sides, J., Tesler M., & Vavreck, L. (2016). The electoral landscape of 2016. *Annals of the American Academy of Political and Social Science, 667*, 50–71.

Sides, J., Tesler M., & Vavreck, L. (2018). Hunting where the duck are: Activating support for Donald Trump in the 2016 Republican primary. *Journal of Elections, Public Opinion and Parties, 28*(2), 135–156.

Tolbert, C. J., Redlawsk, D. P., & Gracey, K. J. (2018). Racial attitudes and emotional responses to the 2016 Republican candidates. *Journal of Elections, Public Opinion and Parties, 28*(2), 245–262.

Turney, S., Levy, F., Citrin, J., & O'Brian, N. (2017). Waiting for Trump: The move to the right of white working-class men 1968–2016. *California Journal of Politics & Policy*. Berkeley: Institute of Governmental Studies. Retrieved from https://escholarship.org/uc/item/1cq9k81z on March 15, 2018.

Wright, W. (2018). This Kentucky school district just voted to let teachers carry concealed guns. *Lexington Herald-Leader*. Available online at: http://www.kentucky.com/news/state/article202341909.html. Accessed March 3, 2018.

Chapter 6
Conclusion

Waking up after the long election night of 2000, and not knowing that they would soon deal with an unprecedented month-long continuation of the contest to determine who would be the next President of the United States, Americans found themselves confronting what seemed like a new political geography and, with it, a new conception of their country. The United States appeared to be divided into two Americas, one a diverse, cosmopolitan, America located primarily on the coasts that favored the Democratic Party, and the other, less diverse, committed to traditional values, and located primarily in the interior of the country that favored the Republican Party. This seemed, to American voters, to be a new reality, and perhaps one that they increasingly felt as the country appeared to divide along cultural dimensions that citizens lived with every day in their lives.

Popular treatments of the supposed red state-blue state divide begin with election night in 2000. But presidential elections didn't begin with the Bush vs. Gore election in 2000. In fact, at that point we had more than 170 years of experience with mass voting in American presidential elections. Our goal in this book has been to provide new insights into our current politics by placing these politics in historical perspective. In doing so, we've illuminated several features of our contemporary politics that can better help us understand the country in which we live.

First, there's nothing unusual about our current era, nor for that matter, anything particularly polarized from a geographic perspective. The red state-blue state maps actually do more to distort our understanding of political allegiances in voting than they do to illuminate them. This is so, first, because states are not homogeneous, monolithic entities. As Barack Obama famously said in his 2004 Democratic

D. Darmofal, R. Strickler, *Demography, Politics, and Partisan Polarization in the United States, 1828–2016*, Spatial Demography Book Series 2,
https://doi.org/10.1007/978-3-030-04001-7_6

National Convention keynote speech, "We worship an awesome God in the blue states, and we don't like federal agents poking around our libraries in the red states. We coach little league in the blue states and, yes, we've got some gay friends in the red states."[1]

Our data in this book are not individual-level data, and thus we can't speak at the level of the individual as Obama did in his famous speech. But what we've uncovered in this book is consistent with the political and cultural heterogeneity within states that Obama identified. States, as containers for heterogeneous preferences, are not the best entities for understanding the political geography of the United States. There are Democratic counties in red South Carolina that vote more like counties in blue California than most observers would expect. And likewise with Republican counties in blue California that are more similar to South Carolina than pundits would expect.

States will always be important entities for elections in the United States given the Electoral College. This is obvious, for example, from the continued discussions of the Clinton campaign's tactics in Michigan, Pennsylvania, and Wisconsin. But states are less helpful for understanding the political geography of voting in the United States. There is simply too much within-state variation in partisan voting for states to occupy the central focus of electoral analyses that they have since 2000. It's time, instead, to look at the sub-state level to understand partisan voting in the United States.

Identifying the political geography of presidential voting at the local level is critical to understanding elections in America. But in order to identify this geography, we need to move beyond mere "eyeball" analyses of partisan voting. The human eye (and mind) has a tendency to identify patterns where none exist, a phenomenon known as apophenia. Thus we can't just look at maps of partisan voting and conclude that there is a spatial structure to this voting. We need to bring diagnostic tools to this question. This book has done this, and in the process has, for the first time, identified the spatial structure of presidential voting at the county level in all elections since the advent of mass voter participation in 1828.

By employing spatial diagnostics, we have been able to identify where regimes of Democratic and National Republican/Whig/Republican strength have been located in each presidential election from 1828 to 2016. In doing so, we have uncovered multiple insights that shed important light on our electoral past—and in doing so, on our present and our future. First, partisan voting is not spatially random. And it hasn't been in any election since mass participation began in earnest with the Jackson-Adams contest of 1828. Instead, there has been a strong positive spatial structuring to partisan voting in the U.S., meaning that counties in the United States have tended to exhibit similar partisan voting as their neighboring counties. Place is, in other words, important for explaining why Americans have voted as they have.

[1]http://www.washingtonpost.com/wp-dyn/articles/A19751-2004Jul27.html. Accessed on March 3, 2018.

Moreover, the political geography of voting has generally not been structured at the level of the state. Substantial variation in partisan voting has occurred within states in each presidential election from 1828 to the present. The Jim Crow era of the Solid Democratic South was, of course, an example of a strong state-level structuring of partisan voting. But this was more the exception than the rule when one considers the full spatial and temporal span of mass voter participation in the United States.

The political geography of partisan voting that we have uncovered has, moreover, not changed in a haphazard manner from election to election. There are, of course, exceptions in which a map bears little resemblance to those that came before or after it—the 1964 and 1972 maps are good examples of this. But again, they are the exception rather than the rule. Instead, the rule has been one of gradual changes over time that reflect gradual changes in the parties' geographic bases.

The results of the spatial analyses also indicate that we haven't become a more geographically polarized nation in recent years. To be sure, Trump's average margin of victory in 2016 was unusually large by Republican standards, the largest such average margin for a Republican candidate since Abraham Lincoln in 1864. This average margin of 61.95 points was, however, exceeded on the Democratic side by all four of Franklin Roosevelt's victories (and was dwarfed by his first three victories). It was also exceeded in multiple elections during the period of the Solid Democratic South.

Equally important, there is little evidence of heightened margins on the Democratic side. Hillary Clinton's average margin of victory of 10.3 points in the blue counties in Fig. 2.50 was dwarfed by most previous average Democratic margins of victory and was smaller than either of Obama's two election victories.

In fact, one can only make the argument that our current political geography is especially polarized if one solely focuses on the past few decades. This is what Bishop (2008) does with the *The Big Sort*, using 1976 as a baseline for analysis. In Chap. 3, we discuss a litany of research that suggests that, while counties may have become more homogenous in partisan voting since this baseline, it is likely not due to lifestyle-based migration choices as Bishop claims. It is more likely that political change has been driven by partisan conversion or generational replacement. It is also important to emphasize the foundational role racial identity has played both in migration decisions specifically and partisan identification more generally. Moreover, 1976 was an unusual year of partisan detente. Taking a historical view, partisans are no more likely to live in "polarized" communities in recent decades than they were during the advent of Jacksonian democracy, or during the Gilded Age. Even the 2016 election, where the percentage of voters living in "landslide" partisan counties reached a relative apogee, has precedent. And it is too early to tell if this unusual election will precede a long-lasting era of heightened geographic polarization, or if it is just that—unusual.

If polarization cannot help us explain current politics as much as we thought, how best can we understand our political maps? We modeled the sources of political geography in the United States since 1828 using a set of common covariates that tap major political faultines in the country: race, immigration, and the rural-urban divide. We found that racial demographics have played a consequential, and time-varying role, in partisan voting. Immigration also became an important influence as levels of immigration to the United States rose. And population density has consistently shaped partisan voting in the United States since the 1920s.

Donald Trump has argued that his presidency reflects the presidency of Andrew Jackson. The contest of 1828 between a populist outsider representing the forgotten man and the consummate Washington insider bears a close resemblance to the election of 2016. While only time will tell if these two presidencies are similar, our analysis does suggest that Trump's candidacy connected to the important, and often overlooked Jacksonian tradition in American politics. We found that the 1828 election is important for understanding the 2016 election. The relationship between county-level voting in the 1828 election and in the 2016 election is stronger than for any election but one (1856) up to 1968. In other words, one could better predict how a county voted in 2016 based on how it voted in 1828 than on how it voted in 1928, or in FDR's four landslide victories, or in the Kennedy-Nixon contest of 1960. For that matter, one can better predict 2016 voting from 1828 voting than from Bishop's baseline Big Sort election of 1976. This is a remarkable fact. Donald Trump appears to have tapped into the zeitgeist of the presidential election from 188 years before in his successful 2016 campaign.

This correlation between the 1828 and 2016 elections gives us a new way to think about our current politics. Too often, popular observers lack historical context in their analyses of contemporary politics. This is reflected in the extremely short time frame from 2000 to today of the popular red state-blue state discussion. The political science research thus far on 2016 shows that objective economic distress did not predict voting in the way that identity politics did. It is important, however, to recognize that this identity politics is not new, but instead taps into one of the principal political traditions in American politics, the folk community of the Jacksonian tradition. If we just focus on current identity politics in trying to explain our current politics, we are missing a valuable historical lens in which to place these identity politics. The 2016 election was part of a tapestry of nearly 200 years of mass voter participation in the United States. By placing this election in historical context, and connecting it up with the 1828 Jacksonian election, we have shown how an often overlooked political tradition resonated in the 2016 election. As William Faulkner (1951/2012, 73) famously wrote, "The past is never dead. It's not even past." Understanding how previous politics shapes our current politics is essential for understanding our current moment.

6.1 Predicting the Future of Political Geography

By placing the geography of partisan voting in historical context, we can focus and temper our predictions for the future. Areas that have been Republican and Democratic strongholds in the Sixth Party system, both within and across states, will not flip over the course of one or two elections. Moreover, the factors that have influenced spatial characteristics of voting—race, immigration, and the urban/rural divide—will likely influence elections well into the future. Given the strength of these factors, though, we can speculate on the change we may see starting in 2020 and beyond.

For instance, the "solid South" may cease to be as solidly Republican as it has been in recent elections. There are many areas in the South that have seen dramatic increases in their urban population centers. For instance, the population in Texas has grown nearly 50% over the past two decades, and much of this increase has been concentrated in areas such as Houston, Dallas, and Austin.[2] Moreover, the Southern region overall is becoming more demographically diverse; according to the U.S. Census, the percentage of the white, non-Hispanic population has shrunk from 66% in 2000 to 58% in 2016. As shown in Fig. 2.50, these trends have already produced regions of Democratic support in the mid-Atlantic, the Mississippi Delta, southern Florida, and southern Texas. In recent elections, they have also made Florida and North Carolina competitive, and they have made Virginia a Democratic-leaning state. Given these trends, as well as the consistent historical role race and population density play in structuring partisan geography, it is reasonable to assume that regions of Democratic support in the South will only increase. Eventually, they may grow formidable enough to turn "red" states in the South, such as Texas or Georgia, purple or even blue. While the South has been "solid" since Reconstruction, the region as a whole may be competitive in the future. This will not come through an abrupt change in one election, though; it will be due to a gradual process of demographic change that has been and will continue to occur over decades.

Furthermore, in Chaps. 3 and 4 we discussed the role the urban/rural divide has played in structuring partisan geography; Fig. 3.4 shows that 70.65% of all counties in the U.S.—almost exclusively rural counties—were "landslide" counties for Trump. The urban/rural divide, though, has not been an historical constant; population density has only had an influence on partisan geography since the 1920s. And if partisan coalitions and cleavages change, the influence of the urban/rural divide may wane. In 2018, both political parties are engaging in internal debate on what it means to be "Democrat" or "Republican." Democrats are wrestling with whether to double down on their base of non-white and professional class voters, or to try to increase their appeal to white working-class and rural voters. Republicans, meanwhile, are wrestling with whether to double down on Trump's base of white working class and rural voters, or to broaden the demographic diversity of their base. If either Democrats or Republicans attempt to broaden their coalition of support, and if they are successful, both urban areas and "flyover" country may not be as politically homogenous as we have seen in recent elections. With this, more of the geographic area of the country would thus be competitive, and the number of voters

[2]https://www.tsl.texas.gov/ref/abouttx/census.html; https://www.census.gov/newsroom/press-releases/2016/cb16-81.html

living in "landslide" counties would decrease. Again, this will not be a change that will manifest over the course of one election, but a longer-term process that will take hold over many elections.

The transformation of the South into a competitive region, or the waning of the urban/rural divide, will certainly represent a dramatic shift in modern partisan politics. But it will not represent a completely novel departure in U.S. political history. As we demonstrate, it will be part of the ebb and flow of partisan politics and political geography. It will be another iteration of the many changes in political landscape we have seen through our country's history, and it will have roots in many of the same factors that have influenced the political map through the centuries.

References

Bishop, B., with Cushing, R. G. (2008). *The Big Sort: Why the clustering of like-minded Americans is tearing us apart.* Boston: Mariner Books.
Faulkner, W. (1951/2012). *Requiem for a Nun.* New York: Vintage.

MIX
Papier aus verantwortungsvollen Quellen
Paper from responsible sources
FSC® C105338

If you have any concerns about our products,
you can contact us on
ProductSafety@springernature.com

In case Publisher is established outside the EU,
the EU authorized representative is:
Springer Nature Customer Service Center GmbH
Europaplatz 3, 69115 Heidelberg, Germany

Printed by Libri Plureos GmbH
in Hamburg, Germany